The changing face of VR

Pushing the boundaries of experience across multiple industries

Edited by

Jordan Frith
Clemson University
and
Michael Saker
City University London

Series in Sociology
VERNON PRESS

www.vernonpress.com

In the Americas:
Vernon Press
1000 N West Street, Suite 1200,
Wilmington, Delaware 19801
United States

In the rest of the world:
Vernon Press
C/Sancti Espiritu 17,
Malaga, 29006
Spain

Series in Sociology

Library of Congress Control Number: 2022939866

ISBN: 978-1-64889-474-9

Cover design by Vernon Press using photo by Martin Sanchez on Unsplash.

Table of Contents

List of figures and tables

Figures

Tables

Acknowledgments

An edited book is a unique kind of beast. We (Jordan Frith and Michael Saker) have each written our own books, but this was the first edited collection either of us had ever put together. The entire process is so different from writing a monograph, and it requires so many people to do so many great things along the way. Consequently, we want to start this acknowledgment section in a fairly obvious place: the authors who contributed to this collection. The two of us wrote a book proposal, evaluated abstracts, assigned peer reviews, gave individualized feedback, and wrote the intro. But it's the authors themselves and the excellent work they did that make this book what it is. They put up with super specific—maybe overly specific—type-A instructions from Jordan about everything from heading styles to language conventions. And they wrote these chapters, met their deadlines, and produced insightful work during a pandemic when so much else was going on. We cannot express enough thanks for the amazing work all the contributors did for this book, and it's just as much each of theirs as it is ours.

When we dreamed up this book, our ideal goal was to put a bunch of authors from different backgrounds into conversation about the ways Virtual Reality (VR) is emerging across various industries. And through a good amount of luck, we received so many great abstracts and what you're holding in your hands is pretty much the exact collection we dreamed of. The authors in this collection include researchers trained in multiple academic disciplines and practitioners who have extensive applied experience writing and developing for VR. The authors are also from four different continents, and they cover areas of expertise that range from screenwriting to environmental communication to live theatre. We are grateful that, through their experiences and efforts, the authors in this book provide a perspective on VR across industries that extends far past the more typical VR focus on gaming or social media. Our names might be on the cover of this book, but this book is the product of so many different contributors who taught us so much as we worked with them to develop their chapters.

We also want to thank everyone at Vernon Press for making this book possible. We have both worked with multiple publishers, and the Vernon Press experience was one of the smoothest either of us have ever had. Our editors were responsive and helpful, and the guidelines they provided were clear and detailed. This book took a lot of work, but neither of us can imagine how much more work it would have been without the solid guidance from the Vernon Press staff.

Now we want to switch to some brief individual acknowledgments, starting with Jordan.

Jordan's acknowledgments: I want to start by offering my deepest thanks to my brilliant, beautiful wife, Stevie. Stevie dealt with me throughout this process, through the late nights giving feedback, developing style guides for authors, and all the more mundane parts that go into putting together an edited book. She was as patient and understanding as she always is, and it feels bizarre to look back and think about the fact that almost our entire marriage has been during a pandemic. And even more so, the germs of this book trace back to the early days of our marriage, so in a way, this book feels like something we did together, and I can't think of anyone I would ever want more by my side.

And I'll conclude my acknowledgments by first thanking Michael for being an amazing collaborator. We've worked together for years, and it's always a highlight of my day whenever we get to meet on Zoom and talk about VR. I'm very lucky to have found such an excellent, kind collaborator.

I also want to thank my wonderful family. I hate talking about my work, so my family never knows what I'm working on, but they put up with me anyways. And of course... I could never write an acknowledgment section without including my closest collaborators: my three rescue pitbulls, Daisy, Tinkerbell and Peaches. I can thank people until I'm blue in the face, but the fact remains that I don't *actually* write with other humans; I write and edit with my pups, and I can't imagine what this process would be like without them. Daisy, Tinkerbell, and Peaches are the ones who sat with me as I spent countless hours emailing, organizing, editing, and writing. They begged for pets to distract me from the work (well...that's mostly Daisy), and they barked for walks when I need a break to refresh. So yeah...I'm not sure what this book would be if I didn't have my wonderful dogs to pet along the way.

Michael's acknowledgments: First off, I want to offer a huge thanks to Jordan Frith for all the hours he has poured into this book, and for taking the lead when I was bogged down. In many ways, this manuscript is the culmination of a bunch of articles we published together over the past few years about VR and the various impacts this technology might have on a range of things. It is no exaggeration to state this text would not be in the shape it is without you.

A massive thanks goes to my fantastic colleagues at City, University of London, and specifically, my colleagues and collaborators in the Department of Sociology—you know who you are. We have been through a lot in recent months. I get to work closely with some amazing people. I never take this for granted. Here's to new beginnings. A huge thanks goes to my exceptional friends and family for the encouragement and inspiration they always provide. I would also like to offer a special thanks to my wonderful mum, for her generosity, kindness and support. We have spoken about this book on many occasions over the past year or so. You are an endless source of light. It is impossible to put into words how much I appreciate you.

Finally, I would like to thank my beautiful wife, Megan, and my wonderful children, Úna and Elliott for everything that they do. Without them, none of this would be possible. There isn't a day that goes by when I am not grateful for the time we spent together, and the pleasure it gives me watching you grow up. I often think about the significance of this period and the memories we are making. In my mind, these are the moments I will go back to.

Author bios

Dr. Alicia Corts is an associate professor and director of the theatre program at Saint Leo University. An accomplished director and media designer, she has worked both nationally and internationally. An active researcher, Dr. Corts is primarily interested in performance in virtual spaces, specifically how identity and gender issues are regulated by online performance spaces. She is also one of the primary investigators on the Kutiyattam Project, an effort to bring the experience of this ancient Indian performance form into VR. She has been published in *Theatre Journal, Ecumenica, notches,* and *Rebus* as well as a number of book chapters.

Leighton Evans is an Associate Professor of Media Theory at Swansea University. He is the author of *Locative Social Media* (2015), *The Re-emergence of VR* (2018) and the co-author of *Location-based Social Media: Space, Time and Identity* (2017) and *Intergenerational Locative Play: Augmenting Family* (2021).

Jordan Frith is the Pearce Professor of Professional Communication at Clemson University. His primary research focuses on mobile media and infrastructure. He is the author of 3 books and more than 30 peer-reviewed journal articles in a variety of disciplines. His third book – *A Billion Little Pieces: RFID and Infrastructures of Identification* – was published by MIT Press in Spring 2019.

Hannah Greber a Ph.D. candidate at the University of Vienna, Department of Communication Science since March 2020. With Univ.- Prof. Dr. Sophie Lecheler from the University of Vienna and Dr. Loes Aaldering from the Vrije Universiteit Amsterdam as her supervisors, she is working on her dissertation "Immersive Journalism - the Future of the News?".

Mary Le Rouge is Director of Writing at the Cleveland Institute of Music, where she teaches writing classes and manages the writing centre. She is a member of CCCCs and ISHR. She has authored articles published in the SIGDOC proceedings, IVLA book of readings, *Peitho, CDQ,* and the edited collection *Civic Education Across Ever-Changing Societies,* and is co-editor of the forthcoming *Embodied Environmental Risk in Technical Communication,* published by ATTW/ Routledge in 2022. Her research lies at the intersection of the humanities and the sciences, looking for ways to improve communication between experts, policymakers, and the public.

Rafael Leal is a screenwriter, executive producer, and VR creator based in Rio de Janeiro, Brazil. He is a Professor of Screenwriting at the Pontifical Catholic University of Rio de Janeiro and a Ph.D. Candidate in Film at Fluminense

Federal University, researching the Poetics of interactive and immersive narratives, with a DAAD-funded doctoral research period in Germany. His creations include the feature film "Too Soon," (Fox), and TV shows "Queenpin" (CineBrasilTV) and "Jungle Pilot" (NBC Universal), whose development was the subject of the chapter "Transcultural Collaboration in Screenwriting: Jungle Pilot's Case Study," published in the book "Transcultural Screenwriting: Telling Stories for a Global World" (Cambridge Scholars). Since 2019, Rafael has been a member of the Screenwriting Research Network's Executive Council.

Chris 'Topher' Maraffi is an Assistant Professor in the Art and Design department at North Carolina State University. He teaches emerging technology courses for real-time storytelling and visual effects at NC State, and previously taught game studies, animation, and extended reality courses at Florida Atlantic University. His Mitchelville XR Tour research project received a National Endowment for the Humanities Digital Projects for the Public grant and is concerned with ethically representing the history of the Reconstruction-era Gullah Geechee heritage site through interactive media.

Jessie Marchessault is a recent graduate of the MA student in the Media Studies program at Concordia University and is currently working in the videogames industry as a User Research Coordinator. She was a member of Technoculture, Art and Games (TAG) research lab and is currently the lead research assistant for Indie Interfaces: MESI Quebec-GPS Discoverability Engine Project which examines and evaluates indie developer support and discoverability initiatives in Montreal. Her research interests include curation and VR, player experiences and communities of play, and representation in games.

Deepak John Mathew is a Professor and Head of the Department of Design at the Indian Institute of Technology Hyderabad (India). He holds a master's degree in Fine Arts (Graphic Arts) from the M.S. University, Vadodara and has a Ph.D. in Design Education from the Centre for Advanced Studies in Education, M.S. University, and Vadodara. Currently, he is a Professor and Head of the Design Department and involved in setting up the Design Department at IIT Hyderabad.

Eryn Parker is a Ph.D. student at City University of London, in the Sociology Department. Her research focuses on the impact of new media technologies on culture, society and policy

Delwyn Jude Remedios is an Assistant Professor at the Department of Design, Indian Institute of Technology Hyderabad (India). He is an animator and illustrator, with a Master's in Animation Film Design from the National Institute of Design (NID) Ahmedabad, as well as a Bachelor's degree in Mass Media from St. Andrew's College, Mumbai. His student films at NID have received National and International recognition (https://vimeo.com/user35411152). Post NID, he has worked in Mumbai as a Visual Designer in the eLearning industry in areas such as visual storyboards, character designs, infographics, illustration and animation. He teaches Animation, Illustration and Filmmaking at the Department of Design, Indian Institute of Technology (IIT) Hyderabad. Currently, he is pursuing his Ph.D. in VR filmmaking.

Dr. Michał Rzeszewski works at Adam Mickiewicz University in Poznan, where he explores issues related to digital geography and geographic information systems (GIS). His interests include virtual and augmented reality (AR) and the relationship between software, code and space in smart cities. He is the head of the Critical Geography Research Unit, where he develops and evaluates web-based participatory mapping tools.

Michael Saker is a Senior Lecturer in the Department of Sociology, City, University of London. His research focuses on digital media and the datafication of society, with an emphasis on the application of mobile and locative media in daily life. He is the co-author of *Intergenerational Play* (Emerald, 2021) and *Location-Based Social Media* (Palgrave, 2017)

Julia Salles is a researcher and lecturer at Université de Montréal. She holds a PhD in Communication Studies from Université du Québec à Montréal and a MA in Contemporary Art and New Media from Université Paris VIII. She is a film and new media curator and producer and member of the programming team for the Clermont-Ferrand International Short Film Festival since 2015. Her research-creation interests focus on information and communication technologies. She is currently investigating the use of artificial intelligence technologies in interactive documentaries and curatorial aspects of VR.

Max Schleser is a Senior Lecturer in Film and Television and Researcher in the Centre for Transformative Media Technologies (CTMT) at Swinburne University of Technology (Melbourne, Australia). He is a Senior Lecturer in Film and Television and Researcher in the Centre for Transformative Media Technologies (CTMT) at Swinburne University of Technology (Melbourne, Australia), Adobe Education Leader, Founder of the Mobile Innovation Network & Association (www.mina.pro) and Screening Director of the International Mobile Innovation Screening & Festival. Max's research expertise is Immersive Media and Creative Arts 4.0 with a focus on Cinematic VR and interactive filmmaking. His research explores Screen Production, Emerging Media and Smartphone Filmmaking for community engagement, creative transformation and transmedia storytelling.

Dr. Bart Simon is the current director of Milieux and an Associate Professor in the Department of Sociology and Anthropology. His areas of expertise include game studies, science and technology studies, and cultural sociology. His game studies and design research crosses a variety of genres and platforms looking at the relation between game cultures, socio-materiality, and everyday life. Some of his work is represented in journals such as Games and Culture, Game Studies, and Loading. His current research on the materialities of play, indie game scenes, and player-makers is funded by the Social Science and Humanities Council of Canada.

Kailan Sindelar is an assistant professor in the department of English at the University of North Florida. She received her Ph.D. from Clemson University and her dissertation was entitled "Rewilding with AR and VR: Facilitating Care with Photography in Immersive Apps." She is also an affiliate of Clemson University's VR and Nature Lab where she collaborates on several interdisciplinary projects. Her research interests include environmental rhetoric, technical and professional communication, scientific communication, climate change communication, augmented reality, VR, extended reality, critical animal studies, and care ethics.

Preface

David P. Parisi
College of Charleston

"*Virtual Reality* won't merely replace TV. It will eat it alive."
-Arthur C. Clarke, quoted on the back cover of Howard Rheingold's *Virtual Reality* (1991)

This collection arrives at a pivotal moment for VR. As Frith and Saker outline in their introductory chapter of this volume, after several cycles of "hype and disappointment" (p. 1) around the technology dating back to the 1980s, we're now seeing an unprecedented amount of commercial investment in VR across a wide range of industries. While use cases for this latest generation of headsets released in the mid-2010s initially focused on gaming, they gradually began expanding to include the wide range of applications covered in this book, with VR seen as a social space, teaching tool, a potential replacement for retail commerce, a tool for design, a means of artistic creation and consumption, and a new mode of storytelling and performance. And at these early stages of VR's most recent wave, fragmentation in VR hardware landscape presented an obstacle for both creating and consuming VR content. The issue of fragmentation could be lessening, however. Meta's (formerly Facebook) standalone Quest 2 headset, retailing for $300 USD, has effectively emerged as the consumer standard, with roughly 15 million units sold as of June 2022.

Despite these positive developments, VR's future trajectory is still very much an open question. Proponents and enthusiasts often speak of its widespread adoption as a foregone conclusion, with VR headsets forecasted to eventually replace smartphone screens as the standard mode of both consuming traditional media (strapping on a headset to watch a Netflix show, for example) and networked digital communication (where multiuser 3D virtual environments like *Horizon Worlds* supplant traditional social media platforms). Because VR provides a purportedly more embodied and "natural" way of interacting with media, existing media forms, according to this narrative, will be left behind by a kind of mediatic evolution. This story is exemplified by enthusiasts' adoption of the derogatory term "pancake game" to denote non-VR video games, which situates VR as the next step in a progressive media evolution that will more fully capture the senses and project the body into computer-generated worlds. As with Clarke's quote above, from over 30 years ago, the assumption here is that

VR won't emerge as a mere complement to existing media forms but rather that it will rapidly make them all obsolete.

At the same time, there's also a persistent evolutionary promise baked into current-gen hardware, an assurance that each component of the current VR interface will gradually improve: headsets will get lighter, screens will increase in resolution and response rate, haptic cues will grow increasingly complex and target a wider surface of the body, motion and gesture recognition will become more accurate, and VR will even expand the range of senses it simulates to even include taste and smell (and it's worth recalling here that in Ivan Sutherland's foundational "The Ultimate Display," he suggested that it should "serve as many senses as possible"). VR in the future will not only make today's media obsolete; it will also make today's VR obsolete.

These deterministic narratives about the inevitability of VR's proliferation and evolution have been buoyed by its recent popular cultural portrayals as a ubiquitous technology. The film *Ready Player One*, when it arrived in theatres in 2018, provided a cinematic depiction of a future where networked VR is widely adopted, immensely popular, and deeply integrated into our social fabric. Although the film positions VR as central to a corporatized dystopia, enthusiasts routinely point to *Ready Player One* as an aspirational illustration of where the technology will inevitably bring us.

Proponents present the deep integration of VR into the social fabric as desirable because they argue VR's increased uptake will lead to broad-based social progress. From its earliest days, more than three decades ago, VR has often been presented in these utopian terms. But the current marketing push has expanded the range of VR's supposed transformative powers. Some of these claims are echoed in this volume, albeit in more muted tones. VR can function as "the ultimate empathy machine" (Milk, cited in Maraffi, this volume, p. 22), allowing us to inhabit the bodies of others, gaining a greater understanding of their embodied experiences, and thereby helping to decrease racist, sexist, and ableist attitudes. With its purported superiority to previous modes of communication, it can serve as a suitable substitute for face-to-face interaction, ameliorating the feelings of loneliness and disconnection that still plague remote communication. Through its "quasi-physicality" (Evans and Rzeszewski, this volume, p. 144), VR can provide a functional replacement for physical space, with embodied spatial cues sufficiently cementing associations between memories and computer-generated environments. It might heighten student engagement in the classroom, leading to better learning outcomes and broader access to quality education, as promised in Meta's recent announcement that it has begun partnering with colleges to turn their campuses in "metaversities. It has the potential to increase feelings of credibility and objectivity in journalism (see Greber, this volume). It will motivate us to exercise, with its immersive character allowing it to succeed where other attempts to push 'exergaming' have failed. At this stage in its hype cycle, it's difficult to imagine a challenge VR proponents *haven't* offered to solve with a sprinkling of metaverse dust.

The problem with this sort of crude technological solutionism—where we turn to technology to solve problems that are social at their core—concerns its deferral of responsibility: if we're waiting for the technology's increased proliferation and rapid maturation, we're hitching our hopes for social progress to corporatized wagons, repeating the mistakes with Web 3.0 that we made with Web 2.0, while also obscuring the possibility that VR might exacerbate rather than address this wide range of social issues. Commendably, while the authors in this volume excitedly explore the new possibilities VR opens up, they do so without resorting to deterministic and techno-utopian narratives pushed by marketers and profiteers. Confronting VR requires us to treat it as a technology without an inevitable and fixed trajectory, one that manifests differently depending on the sociocultural and economic context in which it's deployed. The question of "what can and will VR do?" is best answered by looking at what VR *is* now doing in practice: how VR is being actively used by practitioners and how that use is prompting them to revise and rethink and reconceptualize their existing practices.

By including contributors working in a range of roles, this collection also speaks effectively to the question of who is doing VR development and the challenges VR poses to established modes of media creation and production. Consequently, this collection effectively presents a pragmatic response to this challenge. Sampling broadly from applications beyond gaming, the chapters provide a comprehensive picture of VR—simultaneously as a platform, as a medium, as a technology, as an aesthetic orientation, as an authoring tool, as a genre, and as a business model—that expand our understanding of how thoroughly VR could work across existing communicative practices.

In their introduction to this volume, Frith and Saker repeatedly invoke the idea of VR's imminent arrival, noting that it would, at present, "still be a stretch to say that VR has gone fully mainstream" and then (I think quite justifiably) claiming that "the technology is closer than it ever has been before." (16) This raises another pressing methodological: what criteria we will use to evaluate VR's success or failure? What threshold will VR have to cross in order for us to say it's "gone fully mainstream"? If VR is getting "closer," from whose perspective? Is it a question of solving the many technical challenges that continue to impede its proliferation (including poor graphical fidelity, lack of realistic haptics, ergonomic issues owing to the bulk and weight of headsets, and the ongoing issue of simulation sickness)? Or does VR have to demonstrate a broad-based cultural appeal that we have yet to see evidence of? And how do we speak to these questions without simply parroting the inflated claims of well-funded VR propagandists with a vested interest in its success, or conversely, without dismissing the experiences of enthusiasts who've already had meaningful and transformative experiences enabled by VR interfaces? In short, the open question for us is an evaluative and methodological one: at what point do we start assessing VR based on its actual impact rather than its potential—or virtual—effects?

This all loops us around to the value of the perspectives brought together in this collection. If VR continues to grow, the chapters that follow will be read as previews of a technocultural future where VR is broadly diffused into a range of practices, reshaping them to accommodate this new mode of communication. If VR stagnates, however—if Meta abandons its recently-announced multibillion dollar investments into the technology, if VR contracts back to a narrower focus on gaming applications, or if it merely remains a niche technology with a dedicated core of enthusiasts but not much else–these chapters will read instead as indexes of futures never manifested, snapshots of the technology in a moment of enthusiastic and creative expansion. These practice-based approaches detail the opportunities presented by the new medium while also showing how VR challenges our existing vocabularies and frameworks of creative practice.

Chapter 1

Introduction: The winding road of Virtual Reality (VR) development

Jordan Frith

Clemson University

Michael Saker

City University London

Abstract

This introductory chapter examines the long history of Virtual Reality (VR) and its potentially bright future. The chapter begins by discussing major corporations' significant investment in VR. The authors argue that this significant investment will likely expand VR to more and more industries, which makes the case studies found throughout this book even more relevant. The chapter then details the long history of hype and disappointment about VR before examining its "Renaissance" in the mid-2010s. The chapter then describes the structure of this edited collection and the wide range of topics covered in the book and concludes by arguing that the future of VR is brighter than it ever has been.

***

Introduction

We began soliciting chapters for this book in the fall of 2020 at a point when Virtual Reality (VR) was clearly on an upswing. That fall, Facebook released the Oculus Quest 2—an updated version of its wireless predecessor—which was an immediate hit. And, of course, we were over half a year into the COVID-19 pandemic. For the most part, live events and travel had come to a screeching halt, and in some cases, VR designers had designed platforms to partially fill the void left by travel restrictions. Consequently, the moment seemed ideal to bring together a diverse group of academics and practitioners doing research on disparate forms of VR. VR's popularity was growing, and everything from attending concerts through headsets to wandering through virtual museums seemed more relevant than ever before.

We are now writing this introduction in the early spring of 2022 and the topics covered in these chapters have become even more relevant than we could have imagined. In October 2021, Facebook announced they were rebranding as "Meta," which is the new name for their umbrella corporation (similar to Alphabet as the umbrella company for Google and its offshoots). This major announcement followed weeks of negative press about Facebook, which included evidence of the role Facebook groups played in organizing the January 6th uprising in Washington, DC, alongside whistle-blower revelations that Facebook knowingly uses algorithms on Instagram that are detrimental to users' mental health (mainly to drive ad revenue) (Romo, 2021; Timberg, Dwoskin, and Albergotti, 2021). As a result of the timing, many commentators greeted the major Meta rebranding with scepticism and argued the timing was a cynical attempt to change the public narrative (Fenlon, 2021). However, regardless of the reasons, the Meta rebranding showed that one of the largest companies in the world envisioned a bright future for VR that makes the use cases examined in these chapters even more relevant.

The core of the Meta rebranding was Mark Zuckerberg's belief that the future of digital media will be found in VR and—to a slightly lesser extent—augmented reality (AR). As part of the "launch" event, Zuckerberg announced the company was spending $10 billion in 2021 on building their VR "Metaverse," with similar investments in coming years (Kastrenakes and Heath, 2021). He also announced they were hiring thousands of additional employees to build their VR platforms. And what was maybe most astonishing—and arguably alarming—about the announcement event was the *scope* of the vision for the future of VR. The Meta home page has videos touting the use of VR for everything from exercise to education to entertainment. Zuckerberg himself hyped the supposed "Metaverse" as the "next evolution in social connection" (Meta, 2021, n.p.).

Equally importantly, Meta was far from the only company announcing interest in VR. Companies ranging from Roblox to Sony to Microsoft also announced interest in building a supposed Metaverse, and VR is central to much of that vision. At this point, there's no clear outline of just what a Metaverse would look like and no way of knowing which companies would succeed in building a Metaverse. And honestly, we are not sure that it matters all that much for the future of VR and particularly for the relevance of the chapters in this collection. While the term "Metaverse" may end up being a buzzword people forget about within a few years, the investment in VR will still help the media form grow in ways these chapters document in detail.

The chapters in this book were mostly written before Meta's massive investment in VR and companies ranging from Roblox to Microsoft's announcements of increased interest in the technology, but all the chapters analyse how VR has been used across different settings in disparate ways. VR is often discussed mainly in terms of gaming and social networking, and those are two of the dominant uses. But as this collection demonstrates, VR has also been adapted to many different situations, ranging from live music events to environmental education to museum exhibits, to name but a few. As we explain later, VR adoption was already on an upswing well before Facebook announced their rebrand and their multibillion-USD investment in the future of VR. The technology had already been explored across cultural and educational settings, and if even a small fraction of visions of the "Metaverse"—whether we're talking Zuckerberg's vision or the other companies determined to supposedly build a Metaverse—comes to pass, we can expect to see growth in the use of VR across various industries. Consequently, this book is one of the first to capture the scope of just how differently VR is being used today and begin theorizing the impacts of a broad range of examples across multiple settings in various cultural contexts. We expect the topics covered in these chapters to only become more relevant in the coming years.

The rest of this chapter dives deeper into the world of VR to help situate the case studies and theorizations found in the subsequent chapters. We mainly focus on the history of VR, especially its re-emergence in the mid-2010s. We consciously avoid delving into theories of VR usage for two primary reasons. First, the authors in this collection draw on different frameworks to situate their analyses. Secondly, we want to avoid positing a "correct" way to understand the theoretical dimensions of the media form. But as this chapter shows, the history of VR is important for understanding the case studies found throughout this book. We then follow that history by explaining the structure of the book and discussing the wide array of topics covered by the authors in this collection. Our conclusion then looks to the future of VR, a future that is brighter than it ever has been and filled with the potential to impact more and more practices of everyday life.

The long and often unfulfilled promise of VR

VR occupies a somewhat bizarre place in the digital media ecosystem. On the one hand, VR is an emerging, cutting-edge technology. New devices with new capabilities have been developed in the last half decade, and as discussed above, one of the most powerful companies in the world views VR as the future of digital media; additionally, other major companies like Sony and HTC are also heavily invested in VR. In terms of hype, VR is currently one of the most hyped of all emerging media forms and seems to be on the brink of widespread adoption, especially as 2020's Quest 2 device quickly became one of the best-selling VR headsets of all time. Consequently, if looked at from one angle, VR seems shiny and new.

The irony about positioning VR as "emerging" or as a "bleeding edge" technology, however, is that VR is quite old. VR is far older than, say, social media platforms and significantly older than the changes that ushered in what many people call "Web 2.0." What is often considered the very first VR Head-Mounted Display (HMD)—invented in 1968 by Ivan Sutherland—is about as old as email. And even by the 1980s, VR had entered the popular imagination through influential novels like William Gibson's (1984) *Neuromancer*, and research labs had already been formed to explore VR. Consequently, the idea of VR as "emerging" is a matter of perspective. While it is correct to note that VR may finally be on the brink of hitting mainstream adoption and companies are now investing tens of billions of USD into developing the technology, tech sources were talking about VR as the "future" before most people even had dial-up internet in their homes (Negroponte, 1993). VR, paradoxically, is simultaneously old and new, established and emerging. The history of VR is a history of hype and disappointment, and we are finally at a point where much of that hype might be realized.

Of course, it took many decades for VR to get to the point where the chapters in this book would even be possible. Consequently, examining VR's long history of hype and disappointment is an important context for understanding why the media form is now finding its way into everything from museum exhibits to environmental communication to classrooms. The history is particularly important because, despite the long and winding road of VR development, much of the discourse about the technology—namely that it can be a pseudo "replacement" for the physical world—still survives today. Consequently, while a full VR history is outside the scope of the chapter (see Evans, 2018 for a longer history), the next subsection provides a brief history that we argue is important for understanding both the current VR moment documented in these chapters and VR's possible futures.

The halcyon days of VR

Almost any discussion of the history of a technology must make arbitrary decisions about where to start, and that is certainly true of VR. For example, Moser and MacLeod (1996) argue that VR technology is a continuum of developments that dates back to the sensorial experiences of the panorama in 19th century England. But we choose to start at a more conventional place: Ivan Sutherland's creation of the first headset that could be used to access computer screens. Sutherland's (1965) influential article "The Ultimate Display" "described how one day, the computer would provide a window into virtual worlds" (Vince, 2004, p. 4). Sutherland's vision of people entering virtual worlds through

Head-Mounted Displays (HMDs) was far ahead of its time and remains what many point to as the "beginning" of VR as we know it today.

While Sutherland's work was visionary, his academic publications (as with most academic publications) did not break into the mainstream popular consciousness. Instead, as is often the case, it was science fiction that helped popularize the idea of VR as a transformative technology. William Gibson's 1984 novel *Neuromancer* created a world where people could live their lives in a virtual reality environment called "the matrix" (which inspired the classic 1999 movie *The Matrix*). The novel also famously coined the term "cyberspace," which became one of the main metaphors people used to refer to the Internet more generally throughout the 1990s and early 2000s (and even now in some US presidential debates). Consequently, as scholars have argued, the fact that Gibson's "cyberspace" that superseded the physical world became the dominant metaphor for the early Internet likely shaped much of the 1990s scholarship and popular discussion of the Internet as somehow separate or opposed to the physical world (Baym, 2015; Frith, 2015). In other words, long before VR became commercially viable, it was already metaphorically shaping how people understood the impact of digital media.

VR in the 1980s and early 1990s was also about more than just science fiction imaginaries. During this period, significant research labs like VPL were founded that focused on developing VR, the term "Virtual reality" was popularized by Jaron Lanier, and a few VR devices were even released (Evans, 2018). That hype then became even more pronounced in the 1990s when major companies released early VR technologies (e.g., the Nintendo Virtual Boy), and some futurists labelled VR as the future of the nascent Internet. Nicholas Negroponte (1993), for example, predicted that $25 VR headsets would be available by the mid-1990s and that by 1998 one in ten people would be wearing them on buses or airplanes. And many of Negroponte's contemporaries echoed similar sentiments, with primary sources from tech and academic publications in the late 1980s and early 1990s predicting a near-future dominated by VR headsets (Biocca and Levy, 1995; Jacobson, 1993).

The late 1980s through the mid-1990s were, therefore, what Evans (2018) describes as the "halcyon days of VR culture" (p. 27). Popular sources talked about VR like it was on the brink of becoming a viable, widely adopted technology. Of course, that future did not happen, or at least not in the revolutionary way it was imagined. The cheaper systems were disappointing, while the more expensive systems—ranging from $10,000-$50,000—cost far too much for widespread adoption and were still severely limited (Evans, 2018). Consequently, the halcyon days of VR came up against the limits of technology in ways that differ from some other emerging media forms. As Evans (2018) argues, many media forms face what he calls "cultural lag" in which the imaginaries of how to use a new technology lag well behind the speed at which the technology develops. For example, text messaging as a primary use for a mobile phone only happened years after the feature was available. VR, on the other hand, was a media form with a well-formed popular imaginary—shaped both by media narratives and influential science fiction like *Neuromancer* and Neal Stephenson's (1992) *Snow Crash*—that suffered from what Evans called "technology lag." People had an idea of what VR *could be*, but the technological capabilities weren't close to making those ideas a reality.

The technological limitations put a major damper on the 1990s VR hype. By the 2000s, VR had lost its lustre and was rarely discussed as a major part of the future of digital media. VR obviously did not disappear completely. Instead, the technology was mostly used in specialized forms for purposes like flight simulators and medical training. Almost no notable, widely available commercial VR devices were released in the 2000s or early 2010s, and while HMDs did still sometimes pop up in science fiction, commercially available VR development remained mostly stagnant during this period. If, as Evans argues, the mid-80s through the late 90s were the halcyon days of VR culture, the 2000s through the early 2010s were arguably its dark ages.

A Kickstarter campaign and the revival of VR

By 2011, VR had lost its shine, and while some developers continued working on the technology, the popular excitement once surrounding it was mostly gone. At that point, a prediction that in ten years an edited collection would be able to document the many ways VR has begun to impact a variety of industries would have likely been brushed off as just one more example in the long line of misplaced VR hype. VR's fortunes changed dramatically in the mid-2010s, however, beginning with—of all things—a Kickstarter fundraising campaign.

In 2012, a small start-up called Oculus began a fundraising campaign using the Kickstarter platform. The goal of the campaign was to raise $250,000 USD to fund the development of the Oculus Rift, marketed as a relatively affordable and powerful VR headset designed for gaming. As the fundraising campaign put it, "There are a number of VR headsets out there, but none that deliver a truly immersive gaming experience. Most products either lack the technical features required for believable immersion or sit at a very high price-point ($20,000+) reserved for the military or scientific community" (Oculus, 2012, n.p.). And the fundraising pitch worked, likely beyond the wildest dreams of the developers. By the time the funding period closed, over 9,500 people had donated a total of $2,437,429 USD to fund the project, almost ten times the initial goal. Clearly, even after a long fallow period, the idea of viable VR HMD available to the public was still capable of ginning up excitement.

The 2012 Oculus Kickstarter was a watershed moment in VR's supposed renaissance. Companies had been working on VR throughout the 2000s, but mostly for specialized purposes. Oculus clearly demonstrated that interest remained in a viable, relatively affordable VR headset. Other companies took notice, and Facebook purchased Oculus in 2014 for over $2 billion USD. By the mid-2010s, consumers went from essentially having no viable, commercially available HMDs to having multiple options to choose from. Somewhat remarkably, in a seven-month period in 2016

- Oculus released the Rift CV1 headset in March, four years after fundraising for the device
- HTC released the Vive VR headset in April
- Sony released the PlayStation VR (PSVR) in October

Within about half a year, consumers went from almost no options for VR to the choice of devices from three major companies. And VR has only continued to grow from there. Both Sony and HTC released second-generation devices and have continued developing new technology. But the most dominant player in the contemporary state of VR has been Oculus, or to put it more accurately, Facebook (both now under the Meta umbrella).

Facebook released multiple devices in the latter half of the 2010s through the Oculus brand, including multiple successors to the original Rift. However, arguably the most significant move Oculus made was to focus on developing their Quest line of devices. Here it helps to go into a bit of detail on the materiality of VR to discuss the current moment and near future of the media form.

The promise of VR has often focused on the freedom to be immersed in virtual worlds (Saker and Frith, 2020). But until the release of the Oculus Quest, there was always somewhat of a contradiction between the rhetorics of "freedom" and the actual design of VR HMDs: namely, all but the lowest-end VR headsets typically had to be connected to high-end gaming computers. Consequently, people using these devices were quite literally tethered to the physical world through cords connected to hardware. This design is still true for most VR devices, including the Oculus Rift series, the Sony PSVR, and the HTC Vive. The connections to computers allowed these devices to render faster at lower latency, but they also tended to remain quite expensive and require expensive equipment. That changed with the release of the Oculus Quest in 2019. The Quest line of devices are standalone HMDs, meaning they do not need to be connected to any other hardware to work. And unlike Oculus' first standalone headset—the Oculus Go—the Quest provided six-degree of movement functionality, meaning it could track the movements of the user just like with more powerful tethered VR systems. Maybe most importantly, the Quest and Quest 2 were far more affordable than most other VR headsets and were priced around $300 USD when they were released.

The original Quest was a success. The device sold well and was a standalone headset that, while being less powerful than VR devices like the Rift or the Sony PSVR, was far more affordable and accessible. Not long after its release, Oculus then released the Oculus Link, which was a cable that could connect the Quest to the PC and make it more powerful. In other words, the Quest both offered the portability and freedom of an all-in-one device and the possibility to make it roughly as powerful as higher-end devices. Oculus then indicated that they would focus more on all-in-one headsets than the more expensive Rift line. The Quest 2 was then released in 2021 at an even lower price point than the original Quest ($299 USD), and that device quickly became the best-selling VR HMD of all time. And as discussed earlier, within a year of the release of the Quest 2, Facebook announced their Meta rebrand and their gigantic bet on the future of VR.

As we have argued, VR has certainly been hyped before and failed to gain widespread adoption. This time, however, the hype does seem more likely to come at least partially true. Standalone headsets are now popular and affordable; even as we draft this chapter, articles are coming out about how, based on download numbers, around 2 million Quest 2 devices were given as Christmas presents in 2021 (Perez, 2022). In addition, multiple large companies have invested in VR, and one of the largest companies in the world—Meta—has positioned VR at the centre of its vision of the future. And even before the Meta announcement, VR had already begun to spread. While most HMDs are marketed as primarily gaming devices, this collection of chapters catalogues how people ranging from musicians to educators to historians to environmental communicators have already begun exploring the unique affordances of the media form. As this brief history has shown, if VR's trajectory continues, we can expect these types of uses to become even more commonplace. Consequently, this collection arrives at an important juncture when we need theories and design approaches to both understand VR's current moment and its future potential. In the next section, we outline the structure of this collection and show how this book brings together practitioners and academics from different disciplines to analyse and theorize the many ways VR is already being used across various industries.

Structure of this book

While there is certainly some overlap between the different chapters in this book, we grouped the chapters into four sections to best capture the diverse uses of VR covered by the authors: (1) VR and storytelling, (2) VR and nature, (3)

VR and performance, (4) VR and exhibitions. To explain the structure, we describe each of the four sections and the different chapters they contain.

Section 1: VR and storytelling

The chapters within this section revolve around storytelling and VR. In other words, this collection of chapters explores how VR is currently reshaping and challenging the contours of storytelling by offering novel ways for the audience to experience emerging media texts. Importantly, the potential for VR storytelling has implications for audiences and practitioners alike.

In chapter 2, 'VR storytelling for social justice and the ethics of playing Black bodies,' Christopher Maraffi examine the various ways VR storytelling can be used to confront and educate audiences on social justice issues and Civil Rights history. Here, Maraffi draws on the role play principles established in Augusto Boal's Theatre of the Oppressed through the "spect-actor" and joker figures as a performative lens to explore the potential of this technology to facilitate empathy. In doing so, Maraffi argues that there is an ethical way of transmuting the "spect-actor" into the body illusion of a protagonist of colour, just as non-player characters (NPCs) designed to exemplify joker figures can be employed to facilitate ethical understandings.

Building upon some of the themes established in the previous chapter, in chapter 3, 'An Embodied poetics: screenwriting for immersive media,' Rafael Leal assesses screenwriting through the lens of VR environments. Drawing on substantial industry experience—both as a screenwriter for more traditional projects and VR-based projects—Leal discusses the body of the spectator and how embodiment should be consciously considered by screenwriters for VR. Ultimately, the chapter questions how the concept of the *embodied script* is developed and situated in the context of film production, while complicating the methodological separation between creation and execution found in typical approaches to screenwriting.

In chapter 4, 'Taking stock: VR in journalism,' Hannah Greber attends to the decade-long phase of experimentation VR has gone through in the field of journalism. Following a critical review of relevant literature, Greber considers the assorted ways VR journalism has been produced, the utility of these productions, their ensuing reception, and whether journalistic VR might contribute to the democratic role of journalism. More specifically, this chapter hinges on four propositions. First, traditional journalistic storytelling should be reconsidered when it comes to VR. Second, there is a distinct lack of viable financing strategies for this emerging practice. Third, while VR implicates a marked level of presence that differentiates this technology from other media, the effects of VR journalism are not as clear-cut as one might assume. And finally, she argues that VR journalism does have the potential to complicate some of the core tenets of journalistic practice.

To conclude this section, the chapter by Delwyn Jude Remedios, Deepak John Mathew, and Max Schleser delves deeper into the storytelling potential of VR by examining whether parallel interactive narratives in VR (PIN VR) might offer new ways for audiences to interact with filmic narratives. Through an innovative practice-based research methodology that focuses on PIN VR, the authors provide a detailed account of the production process and narrative design associated with this nascent media form. In so doing, the chapter implicitly engages with many of the themes introduced in previous chapters.

Section 2: VR and nature

The chapters within this second section explore VR's potential to enable new forms of science communication about nature. This section, therefore, continues to consider the changing contours of VR by examining different applications through a lens of environmentalism. The section also explores the utility of VR as possibly extending into the domain of education, presenting an interesting relationship between both physical and digital spaces.

In chapter 6, 'VR's growing role in establishing environmental empathy and its potential to model complex climate systems,' Mary Le Rouge moves beyond the common association between VR and empathy to consider the potential this technology has to both model climates systems and enable users to help develop possible solutions to mounting environmental problems. In her analysis, Le Rouge demonstrates the novel ways the digitality of VR can be used to elicit tangible changes in concrete space beyond the headset. Most importantly, this chapter offers an intriguing response to surrounding discussions about the phenomenology of this technology and its supposed propensity to distance users from their physical environment.

In chapter 7, 'Designing with care ethics in nature-based VR,' Kailan Sindelar examines nature-based VR through the lens of education and therapy, the impact VR apps might have on representations of wildlife and wilderness, and the connections between users and representations of nature this practice might cultivate. Through an examination of three VR applications that focus on connecting people with nature, Sindelar argues designers should embrace the role of "care ethics" in environmental applications to avoid implicitly embracing the typical split between humans and their environments.

Section 3: VR and performance

The third section considers how the supposed authenticity of concrete, physical spaces can be reimagined in the digital space of VR, while also acknowledging how the unique affordances of these virtual environments might challenge the limitations of the physical world.

In chapter 8, 'Being-there when there is not there: Live performance, social VR and COVID-19,' Leighton Evans and Michal Rzeszewski reflect on the emerging role VR played during the Covid-19 pandemic, when a succession of lockdowns and social distancing measures effectively halted live musical performances across the world. As the authors show, VR provided a possible space where artists could partially replicate the experience of live musical performance. Through an embrace of post-phenomenology, Evans and Rzeszewski explore questions about the nature of performance and presence and how VR live performances enable different types of experiences of liveness and embodiment compared to traditional physical performances. Ultimately, the authors argue for a conceptualization of 'being-there' and the important role that supposed authenticity might have on reception of live performances.

In chapter 9, '*The Under Presents*: The potential of the real in virtual theatre,' Alicia Corts moves beyond discussions of authenticity to consider the role VR might play in expanding upon the immersive possibilities of theatrical productions. Through the lens of three virtual performances—*Carney y Arena*, *Collider*, and *The Under Presents: The Tempest*—Corts draws on Lawrence Halprin's RSVP concept for devised theatre to explore how the virtual and physical bodies of both audience and performers should be considered as performative resources. Corts uses this approach to explore the unique aspects of VR and the fact stories can be told that challenge the limitations of the physical. Corts argues VR might challenge the way audiences perceive the world and their place in it. In other words, then, and as touched on in the previous section, the concerns of these chapters extend beyond the phenomenology of the headset by nudging users to reflect upon themselves and their lived experiences within the physical world.

Section 4: VR and exhibitions

The concluding section of this book focuses on how VR has been used in various exhibitions. These chapters have implications for designers, curators, and audiences and examine possible benefits and challenges for adopting VR within exhibition spaces.

In chapter 10, 'Curating for communal engagement in VR,' Julia Salles, Jessie Marchessault, and Bart Simon examine the various challenges associated with curating VR-based exhibitions. Here, the authors consider the various stages audiences move through when in the process of experiencing a VR exhibit, such as the period just prior to putting the headset on. Drawing on a case study, the authors present the salient aspects of these challenges and how these issues relate to the affordances of VR. Following that contribution, they then present different curational strategies that might tackle the challenges they identified through their research.

Finally, in chapter 11, 'Exploring the phenomenology of VR-based exhibits in art museums,' Eryn Parker and Michael Saker critically investigate the impact VR is having on certain exhibits and the novel viewing experiences VR might enable for audiences. Stemming from an original research project conducted at the Anise Gallery in London, United Kingdom, between June and July 2018, Parker and Saker report on 19 semi-structured interviews with visitors who had just experienced the VR portion of this exhibit. Parker and Saker identify three ways the VR impacted the audience. First, users felt more able to move around the exhibit in a manner that differed from more traditional exhibitions. Second, because of the headsets involved, users required more assistance from guides than they otherwise would, with this additional assistance having both positive and negative effects on their experience. Third, the long-established norm of people-watching in museums was reshaped by VR, with some users feeling more conscious of being observed. As the authors show through their data analysis, the impacts of VR within this exhibit were quite nuanced and presented both challenges and possibilities for future museum adoption of VR headsets.

Conclusion: The (possibly) bright future of VR

The future of VR is likely brighter than it has ever been, and beyond new headsets or corporate announcements, the breadth of the contributions in this collection help show VR's significant future potential. The authors in this collection come from four different contents and include practitioners and academic researchers who reflect the broad, interdisciplinary potential of VR as a still-emerging media form. As of this writing, it would still be a stretch to say that VR has gone fully mainstream, though download figures show the Quest 2 may be getting close (Perez, 2022). But the technology is closer than it has ever been before, which is reflected by adoption numbers, corporate investment, and the diverse use cases documented in this book.

Of course, no future is ever certain, and the history of VR hype detailed earlier shows how that is particularly true for VR as a media form. The Meta announcement showed a massive investment in the future of VR and a vision for VR that includes many of the varied settings captured in this collection. As we finalize this introduction, multiple companies have announced plans to build a supposed Metaverse. And while there is still very little clarity on what this Metaverse would be, regardless of how various companies envision it, the mere push for a Metaverse will likely be a boon for VR and will strengthen the use of VR across a wide range of industries and cultural settings.

Because the future is so difficult to predict, all we can do is look at the current state of VR and make educated guesses. And, not to put too fine a point on it, the current state of VR is dominated by Meta. Meta owns the best-selling headsets and Meta has the resources to invest billions of dollars in VR. That certainly may change in the future, which is why this introduction and the following chapters have avoided going into too much detail on individual devices and instead analysed VR in ways that can be just as applicable regardless of what new devices come out in the coming years or what company, and it will very possibly be no company, succeeds in building a supposed Metaverse. As we mentioned above, none of the content covered in this book is dependent on any single vision of the future of VR. Instead, as these chapters show, VR has already begun to take root in various industries and extend past the scope of gaming, and that growth will likely continue in ways we cannot predict. We hope the excellent work done by the contributors to this book shows the varied potential of VR, and we believe these chapters can play a role in helping people think through the many novel ways VR can be used to do everything from rethinking what live performance can be to how we explore museums. The future of VR is bright, and as this collection shows, the future is about far more than gaming and social networking.

References

Baym, N. (2015) *Personal connections in the digital age.* 2nd ed. Cambridge, MA: Polity Press.

Biocca, M. and Levy, M. (1995) *Communication in the age of virtual reality.* London: Routledge.

Evans, L. (2018) *The re-emergence of virtual reality.* London: Routledge.

Fenlon, W. (2021) *The Metaverse is bulls**t.* Available at: https://www.pcgamer.com/the-metaverse-is-bullshit/ (Accessed: 8 January 2022).

Frith, J. (2015) *Smartphones as locative media.* London, England: Polity Press.

Gibson, W. (1984) *Neuromancer.* New York: Ace Books.

Jacobson, L. (1993) *Homebrew VR.* Available at: https://www.wired.com/1993/01/homebrew-vr/ (Accessed: 11 October 2019).

Kastrenakes, J. and Heath, A. (2021) *Facebook spending $10 billion this year on its Metaverse division.* Available at: https://www.theverge.com/2021/10/25/22745381/facebook-reality-labs-10-billion-metaverse (Accessed: 18 January 2022).

Meta (2021) [Tweet] *The metaverse is the next evolution of social connection.* Available at: https://twitter.com/Meta/status/1453800021275545601 (Accessed: 11 January 2022).

Moser, MA. and MacLeod, D. (1996) *Immersed in technology: Art and virtual environments.* Cambridge: MIT Press.

Negroponte, N. (1993) *Virtual reality: Oxymoron or pleonasm?* Available at: https://www.wired.com/1993/06/negroponte-11/ (Accessed: 11 January 2022).

Oculus (2012) *Oculus Rift: Step into the game.* Available at: https://www.kickstarter.com/projects/1523379957/oculus-rift-step-into-the-game (Accessed:11 January 2022).

Perez, S. (2022) *Meta's Oculus VR companion app gained ~2M downloads since Christmas.* Available at: https://social.techcrunch.com/2022/01/06/meta-had-a-good-holiday-as-its-oculus-vr-companion-app-gained-2m-downloads-since-christmas/ (Accessed: 11 January 2022).

Romo, V. (2021) *Whistleblower's testimony has resurfaced Facebook's Instagram problem.* Available at: https://www.npr.org/2021/10/05/1043194385/whistleblowers-testimony-facebook-Instagram (Accessed: 7 January 2022).

Saker, M. and Frith, J. (2020) Coextensive space: Virtual reality and the developing relationship between the body, the digital and physical space. *Media, Culture & Society* 42(7–8).

Stephenson, N. (1992) *Snow Crash.* New York: Bantam Books.

Sutherland, I. (1965) The ultimate display. In: *Proceedings of the International Federation of Information Processing Congress,* 1965, pp. 506–508.

Timberg, C., Dwoskin, E. and Albergotti, R. (2021) Inside Facebook, Jan. 6 violence fueled anger, regret over missed warning signs. *The Washington Post.* Available at: https://www.washingtonpost.com/technology/2021/10/22/jan-6-capitol-riot-facebook/ (Accessed: 7 January 2022).

Vince, J. (2004) *Introduction to Virtual Reality.* Los Angeles: Springer.

Section 1: VR and storytelling

Chapter 2

VR storytelling for social justice and the ethics of playing Black bodies

Christopher Maraffi

North Carolina State University

Abstract

Using Augusto Boal's Theatre of the Oppressed role play principles of a "spect-actor" and joker figure as a performative lens, I analyse VR storytelling experiences by artists of colour that confront social justice issues and educate on Civil Rights history using a variety of platforms from 360 video to room-scale VR installations. I argue that narrative framing to cast a spect-actor into the body illusion of a protagonist of colour can be done ethically, and role play with NPCs designed as joker figures representing the oppressed community can promote empathic understanding.

Keywords: Ethical VR, Experiential Storytelling, Serious Play, Empathy Machine

***

Introduction

The experiential and performative dynamics of VR have increasingly been used by artists of colour to explore issues of social justice related to racism and Civil Rights. In this chapter, I argue that VR storytelling design principles developed for entertainment have the potential to surpass the impact of traditional documentary cinema when ethically applied to social justice narratives as multimodal serious play. The embodied 360 perspective of VR world simulation can safely transport spectators to an immersive representation of an oppressed community, and narrative framing with body illusions can cast them into the role of a protagonist in that community's struggle against systemic racism.

Innovative VR social justice experiences like *Carne Y Arena* (2017), *I Am a Man* (2018), *Traveling While Black* (2019), and *The March* (2020), employ design principles from cinema, video games, and participatory theatre arts to transport viewers into a serious simulation of another time, place, and community. The VR experience of taking the first-person perspective of an undocumented immigrant crossing the Mexico-US border at night, or as a Black sanitation worker in Memphis during the 1968 Civil Rights protests, will be examined through the performative lens of Augusto Boal's Theatre of the Oppressed "spect-actor" and "joker" roleplay concepts (Boal, 1992). Because the $160-billion video game and VR industries still have a diversity problem, most of these projects were collaborations between artists of colour and white developers, with stated goals of enhancing player immersion and empathy towards the portrayed subjects.

As a white designer collaborating on a similar research project for a Reconstruction-era African American heritage site, Mitchelville XR Tour (Maraffi, 2020), I will discuss how the framing of VR storytelling has shifted from perceptual illusions to empathic theatre. I will also address critics of framing VR as an empathy machine on the technical and ethical issues, including the serious play of Black avatars in social justice simulations. In our virtual tour project, we are using extensive historical documents from the Library of Congress to model the world of Mitchelville's nineteenth century 3D structures and historical figures. But to immerse visitors in the illusion of presence in historic Mitchelville and create an empathic experience with the people who lived there, we are working closely with Gullah-Geechee community leaders, preservationists, and performers to represent their unique culture. We are applying some of Boal's Theatre of the Oppressed system to design protagonist avatar roleplay that will encourage performative participation in a docudramatic simulation, while also employing a joker role to guide spect-actor improvisation through narrative framing that contextualizes the scene in a broader community struggle.

From VR as illusion to empathic through multimodal serious play

Previous work has shown that innovative VR entertainment attractions like The VOID's *Star Wars: Secrets of the Empire* (2017) used stage magic principles to create the illusion of being transported to impossibly scaled worlds populated with characters that display uncanny liveness (Maraffi, 2021). The VOID functioned like a nineteenth-century magic theatre, employing simulation-dissimulation techniques to misdirect a guest's attention away from the VR mechanism to a "hyper reality" experience that had a multimodal tangible aesthetic not possible in other types of screen-based media. It showed the potential of VR to be a new type of theatre of attractions that incorporated performative design principles from cinema, games, and participatory theatre. Such VR experiences display a literal 360 representation of the game design concept of a magic circle as world-building illusions (Stenros, 2012), where the players' first-person perspective is located in the centre of an embodied 3D avatar that acts in the virtual world with life-sized AI-controlled interactive characters. The multimodal play experience reflects theatre arts theorist Edward Gordon Craig's vision of a Theatre of the Future performed by "uber-marionettes" or super-puppets (Craig,

1907; 1919). This chapter will focus on exploring how both these successful VR entertainment techniques of world-building and performative role play are also being applied to edutainment or docudrama simulations of nonfiction Civil Rights histories, allowing participants to witness and, in some cases, act in scenes of oppression and emancipation. I will examine how some immersive VR storytellers of colour are reframing the medium from an entertainment illusion machine to an empathy machine for confronting social justice issues related to racial bias.

Before developer Chris Milk's TED talk on the potential of VR to be the "ultimate empathy machine" (2015), movie critic Roger Ebert similarly framed cinema as a "machine that generates empathy" (2005). But the same thing could be said about live theatrical productions, and a recent study did indicate that watching live theatre can increase empathy and pro-social behaviour in spectators (Rathje, Hackel, and Zaki, 2021). Theatre of the Oppressed founder Augusto Boal considered empathy a powerful emotional relationship created between spectator and protagonist, and "...one of the mechanisms of the dramatic ritual which can be put to good or bad use." (Boal, 1985, p. 181). According to Boal, a bad use in Aristotelian tragedy was to wield empathy as a weapon over spectators by making them passive witnesses to the dramatic action using an invisible fourth wall. In his Forum Theatre system developed in the 1970s, he reversed this power dynamic by casting spectators as "spect-actor" protagonists (Boal 1992), thus removing the fourth wall and shifting their perspective to the centre of the scene. With similar improvisational theatre game dynamics to live-action role play (LARP) in Dungeons and Dragons tabletop games developed in the same time period, including a "joker" role that functioned like a Dungeon Master, Boal's system utilized the roleplay dynamics for confronting oppression instead of fantasy entertainment. Though Gonzalo Frasca suggested that Boal's exercises could be simulated in serious video games to address social and political issues (2004), his ludology approach rejected both fictive narratives and nonfiction histories and did not apply Boal's performative principles to player avatar interaction with non-player characters (NPCs) in computer role-playing games (RPGs).

While there has been some research on adapting Boal's methods in VR as a model for generating future-oriented design fictions (Fisher and Kozubaev, 2020), the scope of this chapter is not to translate Boal's Theatre of Oppression system into VR, but instead to draw from his and other performative principles related to empathic theatre. Even more than screen-based video games, Boal's spect-actor concept fits the embodied first-person perspective of VR avatars, where a representation of the player's head and hands can perform a life-sized character in the centre of a 360-world simulation. Performance studies show that playing a character role requires empathy, because actors "...need to be able to empathize with and advocate for our character, no matter how heinous her behaviour." (Blair, 2015). In Boal's method, the spect-actor forms an emotional connection to the "ethos" or struggles of their character in the *mise-en-scene* staging of a social conflict, while the joker figure explains the scenic "dianoia" or motivating reasons for the struggle in the context of confronting community oppression (Boal, 1992).

Boal's decomposition of the empathy dynamics in a dramatic scene to performative interaction between the spect-actor and joker roles reflects the dual-stage magic principles of simulation-dissimulation (Maraffi, 2021). In Magic Theatre, the magician invites a spectator to willingly volunteer for an acting role in the show, and then applies dissimulation in the form of misdirection and "patter" to frame the illusion as a trick to be solved by the audience. Game developer Chris Hecker considered style vs structure decomposition to be essential to solve "hard interactive problems" (Hecker, 2008), and certainly generating wonder or empathy in an interactive experience is a hard problem. In a VR social justice experience based on Boal's empathic theatre concepts, the player as spect-actor would be invited to perform as an avatar of colour, and the system would provide a joker function through life-sized NPC guides or storytellers that represent an oppressed people, and who frame the scene as an ethical problem to be solved by players. What Boal makes clear in his work is that participatory theatre was the original empathy machine, and that empathy is generated by the spectator's perceived closeness to the protagonist and cognitive engagement with the struggles of a community against systemic oppression.

Boal's Forum Theatre application of the ethos and dianoia principles to his spect-actor and joker roles reflects the multidimensional psychology decomposition of empathy to affective and cognitive types. Psychology studies have associated presence and immersion in VR to a sense of embodiment through body-swap illusions that allow users to "see through another's eyes" (Wiederhold, 2020), and by generating empathic concern and self-other merging through embodied "perspective-taking" (Thériault et al., 2021). The VOID VR experiences created body-swap and other embodiment illusions based on stage magic principles of perceptual and cognitive misdirection to facilitate VR storytelling immersion (Maraffi, 2021), which align with cinema studies on perceptual and narrative immersion as a source of character empathy, such that "...perceptual immersion is primary and immediate, narrative immersion is secondary and conceptual, operating largely through cognitive processes resulting from the temporal development of narrative" (Sunderland, 2019, p. 11).

Other VR psychology studies have explored the multidimensional link between empathy and immersion and how presence and agency in embodied VR is central to the empathic experience (Cummings et al., 2021; Barbot and Kaufman, 2020). Some VR embodied perspective-taking studies have indicated a reduction of racial bias through cooperative role play as a Black avatar with other Black NPCs (Patane et al., 2020), and others suggest embodied VR role play in an avatar designed to counter stereotypes can have a Proteus Effect, decreasing implicit racial bias through positive and non-stereotypical traits displayed by the avatar (Bertrand et al., 2018). Theatre, cinema, and psychology studies have similar multimodal framing of both immersion and empathy in a simulated experience through perceptual illusions and cognitive framing. This multimodal serious play framework provides a lens for analysing and designing VR storytelling experiences that address social justice issues related to racial bias.

Diversity and Black character representation in historic video game worlds

Despite the $160 billion USD video game industry being four times the economic draw of films in 2020, including the VR games sector (Santos, 2020), there is still a diversity problem in who produces and is represented in game titles. Developer interviews indicate there are normative pressures on companies to repeat popular narrative tropes and cast a white male figure as the default player character, a design decision that is often facilitated by casual or colour-blind racism in the industry (Srauy, 2019). Ubisoft's popular *Assassin's Creed* franchise (2007-2020) introduced playable Black protagonists in *Assassin's Creed 3: Liberation* (2012) and *Assassin's Creed 4: Freedom Cry* (2013), both downloadable additions to the main storyline, using the science fiction plot device of time travel to situate players as Black freedom fighters in eighteenth-century New Orleans and Port-au-Prince during the North Atlantic slave trade.

Game studies researchers have argued that diversity in character representation can expand the potentials of play (Murray, 2018), fulfil an ethical obligation to the variety of video game consumers (Neely, 2017), and is a way to foster real-world empathy for others (Gilbert, 2019). Interviews with high school students who played *Freedom Cry* indicated an increased empathy for enslaved peoples by playing Black protagonist Adewale and experiencing how much freedom mattered to the character. *Freedom Cry* has been called "counter-hegemonic commemorative play" which uses character role play "to enable the legitimacy and acceptability of marginalized identities" (Hammar, 2017, p. 388), possibly through a Proteus Effect of performing the Adewale avatar. Murray compares the Aveline de Grandpré character in *Liberation* to the historical figure Harriet Tubman, especially the intersectionality of the "persona" character dynamic, where players must change costumes or perform different social roles to infiltrate plantations and emancipate Black NPCs (Murray, 2018). Though praised extensively by game critics (Narcisse, 2013), *Liberation* is not without criticism, as Ubisoft hired white actress Amber Goldfarb to voice the main protagonist rather than a performer of colour (Narcisse, 2012). This highlights some of the ethical issues of representing bodies of colour for gameplay in a greater entertainment industry that continues to lack diversity and has a history of employing white performers to portray stereotypical ethnic characters in twentieth-century cinema all the way back to nineteenth-century theatrical minstrelsy.

Ubisoft's *Assassin's Creed Liberation* and *Freedom Cry* video games frame their time travel metanarrative with the tagline "History is our playground" and has been criticized for representing enslaved Africans as gameplay resources while sanitizing the extreme abuse of slavery in the game world (Murray, 2018). In addition, though these titles have been shown to increase student engagement with historical periods through role-play, the interaction may not promote critical thinking about how the narratives are framed (Gilbert, 2019). The lack of cognitive engagement with the portrayed history may be a result of lacking a joker figure that represents the oppressed community and comments on broader social themes. Other games like *Mafia 3* (2017) did design some joker dynamics into the Father James NPC, a Black priest and fellow Vietnam War veteran who continuously challenges Lincoln Clay on his violent choices and how they affect the community. The game's extensive use of 1960s diegetic media in the form of staged radio and television broadcasts frequently creates an immersive documentary aesthetic, situating the game world in a simulation of the Jim Crow South in the Civil Rights era. Although gameplay as Clay requires dynamics that resemble the violent Black-man stereotype, performing Clay's violence in the context of believable systemic racism has a racial justice affect that can feel transformative to players (Leonard, 2019).

Virtual green book: VR time travel to explore Black civil rights landmarks

The *Negro Motorist Green Book* was an essential travel guide for Black Americans published by Victor Green from 1966-1967 when the systemic racism of Black Codes and Jim Crow laws made car travel hazardous for people of colour (LOC, 1936). The metaphor of a virtual Green Book seems fitting for VR applications that create the illusion of going back in time to experience safe spaces in the Black community that were central to the struggle for Civil Rights. In this section, I will examine two such VR experiences of historic Green Book landmarks; one, a room-scale VR 3D simulation of the Lorraine Motel during the 1968 Memphis Black sanitation workers strike where Martin Luther King Jr was assassinated, and the other, a 360 VR docudramatic experience of how Ben's Chili Bowl diner in Washington DC sustained the Black community throughout the Civil Rights era. Using a multimodal serious play lens, I will examine whether the illusion of historical world-building using the perceptual immersion of VR may lead to affective empathy, and whether the storytelling approach has potential for narrative immersion to generate cognitive empathy.

Derek Ham's *I Am A Man* (2018) room-scale VR experience affords you the role of a Black sanitation worker during the 1968 Memphis sanitation labour strike that led up to the assassination of Martin Luther King Jr at the Lorraine Motel. Originally created for the Oculus Rift and now available on the Oculus Quest, the experience features several interactive 3D scenes with embedded documentary media, interspersed by Black transition spaces where you see historical photos and hear interview footage by former Memphis sanitation worker Taylor Rogers. Ham's use of embedded documentary footage in VR was intended to enhance immersion and generate an empathic experience: "Facts and photos come alive in VR with the help of music, lighting, and other intricately planned details, creating an 'intense' experience that encourages empathy and connection" (Klocke, 2018). The first interactive scene casts you into the role of a Black sanitation worker on a Memphis city street in the early morning hours. Your first-person perspective reveals an alleyway with trash cans while you hear a Memphis blues soundtrack mixing with diegetic sounds of a beeping trash truck as it backs up to your position. Since the only interaction available is to pick up the trash cans, you proceed to throw trash in the sanitation truck and notice that your virtual hands are that of a Black male. In the next scene, you are presumably sitting at your own kitchen table after a meal, your wife washing dishes in the background, and proceed to pick up and read period newspapers and homemade flyers about the strike. In the corner is a small Black and white television playing news interviews of white Southerners defending segregation. Another scene shows sanitation workers marching down a sidewalk wearing "I Am A Man" signs, with national guard tanks and police cars on the road behind them.

The world-building effect of the *I Am A Man* VR scenes places you in the centre of a historical simulation of 1968 Memphis, with perceptual immersion as a spect-actor in the role of a Black sanitation striker, while narration of a real-life striker in the transitions and other embedded documentary media promotes cognitive immersion. Like *Mafia 3*, the 1960s documentary media aesthetic is affective, but unlike *Mafia 3* my avatar cannot fully join the fight. The limited performative affordances don't quite make you a protagonist in the scenes, such as marching in the strike or interacting directly with any NPCs, including the disembodied voice of Rogers who represents the community. This lack of agency maintains a partial fourth wall separation between your avatar and others in the community, creating an aesthetic of quiet social isolation common to many VR experiences that lack character dialogue. Where the docudramatic simulation does reach a level of empathic theatre is at the Lorrain Motel on April 4th. The scene situates your avatar in the parking lot of the motel with King on the balcony, going to black just before the fatal shot is heard. Then you are on a city street with sounds of rioting and fires burning, standing in front of an electronics store with televisions tuned to President Kennedy addressing the nation about King's murder and urging Black Americans not to respond with violence. Perhaps the lack of agency and social isolation works dramatically in these scenes because of the helplessness your character might have felt when faced with King's assassination. The experience ends with a simulation of King's motel room as it looks today at the site of the National Civil Rights Museum in Memphis, allowing you to pick up objects not accessible in the real exhibit.

The barrier of a virtual fourth wall is even more noticeable in VR films shot with a 360-camera, where the only interaction is looking around from a single point in space, thus constraining your embodied movement as a spect-actor. The 360-perspective still lends itself to perceptual immersion and presence in the virtual world but cannot afford the performative agency of a protagonist in the narrative. The VR film *Traveling While Black* (2019) transports you to Ben's Chili Bowl in Washington DC, a Green Book listed diner in the 1960s and a still vibrant landmark in the Black community today. In the VR film, the diner itself takes on the joker role: representing the community and connecting historical segregation and discrimination to contemporary issues of systemic racism and police violence. Interviews shot in Ben's Chili Bowl, where the VR spect-actor invisibly watches from the counter or in a booth, are interweaved with narrated documentary footage and historical dramatizations. In some interviews, the diner is open and bustling with people, while other scenes where DC resident Sandra Butler-Truesdale describes her childhood segregation are dark and empty, with documentary footage projected on the walls of the diner-like visible memories. Williams has stated that the experience is intended to be a conversation starter about racial issues that "America hasn't really dealt with" (Carey, 2019), and features an interview in the diner with the mother of Tamir Rice, a twelve-year-old Black boy killed by a white Cleveland police officer in 2014. The 360-world and interviews create some affective moments of filmmaking, but the disembodied and constrained interaction highlights some of the drawbacks of using only 360-video for VR storytelling.

Traditional cinematography aesthetics of a disembodied ideal spectator can be problematic in VR because you feel powerless to interact or even be seen by others in the world. Your inclination is to interact in some way because of the immersive perspective, but even moving your body a little bit will shift the entire world, destroying the perceptual 360 illusion and possibly causing motion sickness. In addition, the intimate closeness of the 360-camera to interview subjects in the diner booth, who never acknowledge your presence by looking at you, creates an awkward voyeuristic aesthetic of being a ghost or a fly on the wall. This is particularly unfortunate because Ben's Chili Bowl represents a community that disembodied spectators cannot join. Other VR films have addressed this 360-video limitation by assigning the spectator a silent witness role through scripted actor behaviour towards the camera. In the first episode of Van Jones' *Messy Truth in VR* (2019) series, the spectator is treated as a Black child in the passenger seat of a car when a police officer stops their father while driving. Even though you have no body or ability to act in the 360-video world, the other actors cast you into a child's role that fits the situation and your lack of agency, which affords you some perceptual and narrative immersion in the scene. Jones saw the experience generated empathy, "Even police officers said that it gave them a completely different set of insights into those situations" (Ramos, 2020).

Serious play: Towards an empathic VR theatre of the oppressed

This section will look at two museum installations that incorporated performative spect-actor role play dynamics in a theatrical room-scale VR production where filmmakers of colour collaborated with high-end visual effects companies to create innovative interaction with hyper-real NPCs. Alejandro Iñárritu's *Carne y Arena: Virtually Present, Physically Invisible* (2017) received a special academy award for "a visionary and powerful experience in storytelling" (Tapley, 2017) that used VR to transport visitors to the Sonoran Desert to walk with undocumented immigrants trying to cross the Mexico-US border at night. It "merges artistry, documentary, technology, exhibition design, and magical realism" (Smith, 2019) to safely transport people into a simulation of a very dangerous journey in order to feel empathy for others who have been through it. Iñárritu, an Oscar-winning filmmaker known for films like *Babel* (2006), *Birdman* (2015) and *The Revenant* (2016), collaborated with ILMxLAB to craft a performative VR experience that required visitors to remove their shoes to facilitate perceptual immersion, "It gets into emotional intelligence, which is less intellectual and radical and can see things clearly because the body doesn't lie. Your feet touching the sand has a huge impact; your body's telling you this is true." (Utichi, 2017). In the pre-production process, Iñárritu interviewed over eighty undocumented immigrants and refugees to develop his six-minute narrative, inviting some of them to represent the virtual immigrants (Heliotis, 2017). ILMxLAB scanned their bodies and movements to develop authentic VR characters, even simulating some of the clothes worn in their real-life trek across the desert.

The multi-act experience begins in the "cold room," where the temperature of the space was made uncomfortably cool to mimic an immigrant detainment room, and many shoes and other lost objects found in the desert were staged in the theatrical space (Stearns, 2019). Visitors are asked to remove their own shoes and store them in a locker, a performative act that casts them into the spect-actor role of an immigrant. Visitors then enter a 50x50' VR stage covered in sand and lit with orange lights, creating the ambient aesthetic of a desert landscape at sunset. After being fit with a headset, visitors find themselves alone in a realistic virtual desert and are joined by a group of immigrants with their "coyote" guide. While walking through the desert, US border patrol agents with bright lights, guns and helicopters suddenly confront the group and the situation escalates. The experience begins as a realistic scene, with helicopter wind and the shock of being discovered, and then transitions to magical realism of seeing through bodies to the beating hearts of agents and immigrants, and a floating boat appears on a table symbolizing the journey across the ocean of sand. After the VR scene ends, visitors retrieve their shoes and walk down a hallway that displays physical pieces of the border wall, to a dark room where they are confronted by the real immigrants telling their stories in video interviews. This third act has some features of a many-faced joker persona, relating the spect-actor experience to the broader immigrant community and the social justice issue of policing at the border.

A more recent room-scale VR museum installation, Time's *The March* (2020), also takes a multi-act theatrical approach. The first room in the installation focuses on audio interviews from Black freedom fighters in Civil Rights protests from the *Montgomery Bus Boycott* in the mid-1950s till the 1963 *March on Washington for Jobs and Freedom* protest where Martin Luther King Jr delivered his *I Have a Dream* speech at the National Mall. The narration includes first-hand accounts of roleplay done by Freedom Riders to train for non-violent protests; storytelling of the dangers faced by marchers such as firebombed buses and fire hoses; and ends with the sound of marching feet as Viola Davis invites visitors "to go back" in time to join the march on Washington. Davis' narration serves some of the joker function, and her audio is illustrated by motion graphics that further engage visitors through text animation. The project was a collaboration between Davis, Black filmmaker Alton Glass, Time's Mia Tramz, and the visual effects company Digital Domain, employing the latest character visual effects to create a representation of King that is the "most realistic human performance in VR to date" (Patterson, 2020). The virtual King simulation is featured in the second room, where visitors enter a 10x15' VR stage or "play area" to join in a spatial representation of the 1963 march. The third room in the installation uses documentary media and AI-controlled interview footage of march organizer Joyce Ladner to "help you connect everything you just learned about, back to the present day," including contemporary racial issues of police violence and Black Lives Matter protests.

Though curtailed to a limited run at the DuSable Museum of African American History due to COVID-19, the reviews indicate that the highlight of the experience is when virtual King glances directly at visitors during his speech, acknowledging their presence in the historical simulation. Black actor Stephen Ferguson was performance captured by Digital Domain to recreate King's movements and voice, which animators then refined using archival footage from the speech (Patterson, 2020). In addition, AI and machine learning added hyper-real details to the NPC, even to the point of simulating subtle blood flow skin dynamics in the face when King is orating. Emotional reactions like "it will give you chills that no digital figment should be capable of conjuring" (Johnson, 2020), and "meeting his gaze was more challenging than I'd assumed it would be" (Austin, 2020), shows how even a simple performative interaction with a convincing virtual icon can be affective. Developers also captured over eighty actors in period costumes to create virtual marchers for the scene, and even included a real freedom marcher and his grandson. As spect-actors, visitors walked with the life-sized virtual marchers and stood with them at a representation of the National Mall, and then King as an embodied joker figure representing the 1960s Civil Rights movement, delivered his iconic speech. In the final moments of the experience, Davis' voice returns to challenge visitors to further action in the real world (Johnson, 2020).

Discussion of VR as an empathy machine and the ethics of playing Black bodies

There has been recent criticism of framing VR as an "empathy machine" (Milk, 2015), with both ethical (Nakamura, 2020) and technical (Eitzen, 2021) arguments against it. Nakamura considers VR's modality to be visceral, not cognitive or hermeneutic, primarily a product of white developers to provide "absolution framed as information" (Nakamura, 2020, p. 54), and not artists of colour speaking their own truth through VR storytelling. She considers the perceptual immersion of "virtuous VR" to be a false and toxic form of embodiment resulting in "identity tourism" that exploits suffering, by invading the private space of marginalized people "...as a way of digitally producing feelings of empathy, sadness, and fear that replace social and institutional responsibility and reform." (Nakamura, 2020). On the technical side, Eitzen has made the argument that VR cannot be a good empathy machine because it lacks the framing of cinema that guides spectator attention, such as close-ups and transitions that switch viewpoints, and because interaction transfers attention away from others to your own experience due to the "bottleneck of attention" (Eitzen, 2019). He did note that VR excels at "second-person storytelling" where figures speak directly to you to create a relationship but can't compare to cinema's "third-person" storytelling that creates the "invisibility of the author" through "inattentional blindness," where spectators forget themselves while observing others (Eitzen, 2021).

The first problem with Eitzen's cinematic framing argument is live theatre, which has been making spectators feel for characters for far longer than cinema without the framing techniques of close-ups and switching viewpoints. From classic Aristotelian tragedy to contemporary Broadway shows like Les Misérables that consistently bring audiences to tears regardless of whether performed in a proscenium or arena stage, theatrical productions direct spectator attention in a way that is distinctly different than cinema but still affective. In addition, Eitzen's preferred third-person storytelling creates the very kind of passive spectator empathic affect that Boal challenged with his participatory spect-actor and joker principles. Also, second-person VR storytelling should be expanded to cover performative roleplay as a vehicle for immersion and empathy in theatre, LARP, and RPGs. Performing an avatar role focuses player attention on the character's experience and can lead to investing in their relationships with others in the scene, which helps to prevent ego-oriented distractions and both immersion and flow for character actors (Maraffi, 2021). Social pressures from other actors in the scene, whether living or NPCs, can guide or challenge the protagonist if they go off-script or get distracted. In RPGs, role play ties the player's gameplay to the narrative and playing out of character could create ludo-narrative dissonance (Hocking, 2007), reducing narrative immersion for spect-actors. This highlights another problem with Eitzen's argument, which is framing VR as predominantly a monomodal storytelling experience, while I maintain its potential for spect-actor affect is in the interplay between storytelling modes from theatre, cinema, and games. Switching between gameplay and cinematics in RPGs is a storytelling feature similar to switching vantage points in cinema. Spatial storytelling in room-scale VR is still an area of active research, but because the 360 perspective scales the interaction to be more like live theatre while still looking through a virtual camera, it will undoubtedly be multimodal.

Many of the problematic examples given by Eitzen and Nakamura were from 360-video journalism of marginalized communities, such as Milk's *Clouds Over Sidra* (2015), which shows life in a Syrian refugee camp as narrated by a twelve-year-old girl who lives there. The limited 360-camera interaction without any kind of dramatic staging or embodied role in the virtual space creates ludo-narrative dissonance between the illusion of being transported into a marginalized community, but not given any way to interact with others. Like in *Traveling While Black*, this aesthetic of being an invisible ghost or a fly on the wall is not conducive to a cooperative perspective-taking affect that has been shown to reduce racial bias. Instead, it may create an aesthetic of voyeurism. *Traveling While Black* does show how dramatic staging and visual effects can direct attention in a 360 video, and *The Messy Truth in VR* shows how role play can improve embodied agency and narrative immersion.

In this chapter, I have used Boal's spect-actor and joker principles to develop a serious play design framework for analysing VR experiences by several artists of colour, who used the medium to interpret the stories of marginalized people facing systemic oppression. Not only does this framework address many of the issues raised by Nakamura and Eitzen, but it also aligns with other studies that find embodied and targeted approaches to empathic VR can be ethical (Rueda and Lara, 2020). Using these examples, we can start to develop ethical best practices for designing serious play VR experiences. In both *Carne y Arena* and *The March* installations, a multimodal theatrical structure was used to first cast visitors into spect-actor roles as members of the represented community, followed by a room-scale VR theatrical scene with liveness that was perceptually immersive, and ending with a joker-like documentary media narrative to promote cognitive understanding of the community struggle. *I Am A Man* staged a call-and-response dynamic between alternating perceptual VR scenes and cognitive narrative through documentary media, demonstrating how avatar role play as a member of the oppressed community can be done ethically.

Lack of diversity in video games has shown the need for playable Black protagonists (Srauy, 2019; Murray, 2018), but the history of stereotyping Black bodies and performing minstrelsy in entertainment media make it essential that embodied role play does not become a twenty-first-century Blackface. The perceptual body illusion of seeing your avatar's Black hands in *I Am A Man* aligns with the performative tasks of emptying trash and narrative framing of embedded documentary media of the 1968 strike, to cast you as a protagonist in that historical struggle. *The March* displays the affective power of a joker figure as liveness illusion in the NPC of King and shows that even a simple interaction like making eye contact in the historical context of King's iconic *I have a Dream* speech can be moving. These examples also demonstrate how to ethically produce VR media that interprets the often-suppressed stories of a racially oppressed community, through collaboration of a diverse group of artists and developers who are in close consultation with members of the represented groups, and by hiring people of colour throughout the production process, especially writers, artists, performers, and voice actors, who best represent bodies of colour.

References

Austin, P. L. (2020) 'How TIME Re-created the 1963 March on Washington in VR,' *Time*, 20 Feb. Available at: https://time.com/5784009/martin-luther-king-jr-virtual-reality/ (Accessed: 27 May 2022).

Barbot, B. and Kaufman, J. (2020) 'What makes immersive VR the ultimate empathy machine? Discerning the underlying mechanisms of change.' *Computers in Human Behavior*. Vol111, October, 106431. https://doi.org/10.1016/j.chb.2020.106431 (Accessed: 27 May 2022).

Bertrand, P. Guegan, J. Robieux, L. McCall, C. and Zenasni, F. (2018) 'Learning Empathy Through VR: Multiple Strategies for Training Empathy-Related Abilities Using Body Ownership Illusions in Embodied VR.' *Frontiers in Robotics and AI*, 5:26. https://doi.org/10.3389/frobt.2018.00026 (Accessed: 27 May 2022).

Blair, R. (2015). Notes on Empathy, Cognitive Neuroscience, and Theatre/Education. p-e-r-f-o-r-m-a-n-c-e.org, 2. Available at: http://p-e-r-f-o-r-m-a-n-c-e.org/?p=1101 (Accessed: 27 May 2022).

Boal, A. (1985) *Theatre of the Oppressed*. Originally published in Spanish as Teatro de Oprimido in 1974. Translated by Charles A. and M. L. McBride. Theatre Communications Group. New York.

Boal, A. (1992) *Games for Actors and Non-Actors.* Translated by Adrian Jackson. Taylor and Francis, Routledge, London and New York.

Carey, M. (2019) 'Director Roger Ross Williams on Emmy-Nominated VR Film 'Traveling While Black': "It's A Big Moment",' *Deadline,* 26 August. Available at: https://deadline.com/2019/08/traveling-while-Black-director-roger-ross-williams-emmys-vr-interview-1202704524/ (Accessed: 27 May 2022).

Craig, E. G. (1907) *The Actor and The Uber-Marionette.* The Mask (self-published periodical), Florence.

Craig, E. G. (1919) *The Theatre Advancing.* On Learning Magic. Little, Brown, & Co. Boston.

Cummings, J. Tsay-Vogel, M. Cahill, T. Zhang, L. (2021) 'Effects of immersive storytelling on affective, cognitive, and associative empathy: The mediating role of presence,' New Media & Society. https://doi.org/10.1177/14614448209 86816 (Accessed: 27 May 2022).

Ebert, R. (2005) 'Roger Ebert on Empathy.' Dedication Speech at the Chicago Theatre. Available at: https://www.rogerebert.com/empathy/video-roger-ebert-on-empathy (Accessed: 27 May 2022).

Eitzen, D. (2019) 'Why VR is NOT an empathy machine.' Available at: https://vimeo.com/353650572 (Accessed: 27 May 2022).

Eitzen, D. (2021). The Problems and Potentials of VR for Documentary Storytelling. *Cinergie – Il Cinema E Le Altre Arti,* 10(19), pp. 43–55. https://doi.org/10.6092/issn.2280-9481/12212 (Accessed: 27 May 2022).

Fisher, J. and Kozubaev, S. (2020) 'Applied Theatre for Developing Participatory Design Fictions in VR.' Proceedings of CHI 2020 Extended Abstracts, April 25–30, Honolulu. Available at: https://www.biopolisproject.com/wp-content/uploads/2020/07/Applied-Theater-and-VR-DIS2020.pdf (Accessed: 27 May 2022).

Frasca, G. (2004) 'Videogames of the Oppressed: critical thinking, education, tolerance and other trivial Issues.' Chapter in: *First Person: New Media as Story, Performance and Game,* ed., Pat Harrigan and Noah Wardrip-Fruin, MIT Press. Cambridge.

Gilbert, L. (2019) 'Assassin's Creed reminds us that history is human experience: Students' senses of empathy while playing a narrative video game,' Theory & Research in Social Education, 47:1, pp. 108-137, https://doi.org/10.1080/0093 3104.2018.1560713 (Accessed: 27 May 2022).

Hammar, E.L. (2017) 'Counter-hegemonic commemorative play: marginalized pasts and the politics of memory in the digital game Assassin's Creed: Freedom Cry,' *Rethinking History,* 21:3, pp. 372-395, https://doi.org/10.1080/13642529.2016.1256622 (Accessed: 27 May 2022).

Hecker, C. (2008) 'Structure VS Style.' Game Developers Conference Proceedings Keynote Lecture. San Jose. Available at: https://www.chrishecker.com/Structure_vs_Style (Accessed: 27 May 2022).

Heliotis, P. (2017) 'Filmmaker Alejandro G. Iñárritu on Breaking the Fourth Wall,' *AnOther,* 05 September. Available at: https://www.anothermag.com/art-photography/10124/filmmaker-alejandro-g-inarritu-on-breaking-the-fourth-wall (Accessed: 27 May 2022).

Hocking, C. (2007) Ludonarrative Dissonance in Bioshock: The problem of what the game is about [Blog post]. *Click Nothing: Design from a Long Time*

Ago, Oct 07. Available at: https://clicknothing.typepad.com/click_nothing/2007/10/ludonarrative-d.html (Accessed: 27 May 2022).

Johnson, S. (2020) 'First look: In VR, Martin Luther King Jr. delivers soul-stirring 'Dream' speech at DuSable Museum.' Chicago Tribune, 27 Feb. Available at: https://www.chicagotribune.com/entertainment/museums/ct-ent-martin-luther-king-virtual-reality-the-march-review-ttd-0228-20200227-sji62kswmzasxgsi5rjjvsbqba-story.html (Accessed: 27 May 2022).

Klocke, C. (2018) 'I AM A Man | VR Civil Rights App,' NCSU College of Design, 25 January. Available at: https://design.ncsu.edu/blog/2018/01/25/i-am-a-man-vr-civil-rights-app/ (Accessed: 27 May 2022).

Leonard, D. (2019) 'Virtual Anti-racism: Pleasure, Catharsis, and Hope in Mafia III and Watch Dogs 2.' *Humanity & Society*, Vol. 44(1), pp. 111-130. https://doi.org/10.1177/0160597619835863 (Accessed: 27 May 2022).

LOC. (1936) *The Negro motorist Green-book*. New York City: V.H. Green. [Periodical] Retrieved from the Library of Congress, https://www.loc.gov/item/2016298176/ (Accessed: 27 May 2022).

Maraffi, C. (2020) Historic Mitchelville AR Tour NEH Design Document. Available at: https://tophermaraffi.files.wordpress.com/2020/06/mitchelvilletourdesigndocfin_maraffi_2020.pdf (Accessed: 27 May 2022).

Maraffi, C. (2021). 'Stage Magic as a Performative Design Principle for VR Storytelling.' *Cinergie – Il Cinema E Le Altre Arti*, 10(19), 93–104. https://doi.org/10.6092/issn.2280-9481/12234 (Accessed: 27 May 2022).

Milk, C. (2015) 'How VR can Create the Ultimate Empathy Machine.' TED 2015. Available at: https://www.ted.com/talks/chris_milk_how_virtual_reality_can_create_the_ultimate_empathy_machine (Accessed: 27 May 2022).

Murray, S. (2018) 'Poetics of Form and Politics of Identity; Or, Games as Cultural Palimpsests,' in *On Video Games: The Visual Politics of Race, Gender and Space*. Bloomsbury Academic, pp. 67–113.

Nakamura, L. (2020) 'Feeling good about feeling bad: virtuous VR and the automation of racial empathy,' *Journal of Visual Culture*, 19(1), pp. 47–64, https://doi.org/10.1177/1470412920906259 (Accessed: 27 May 2022).

Narcisse, E. (2012) 'White Actress Will Voice Assassin's Creed's Black Heroine,' *Kotaku*, 29 August. Available at: https://kotaku.com/white-actress-will-voice-assassins-creeds-Black-heroine-5939009 (Accessed: 27 May 2022).

Narcisse, E. (2013) 'This Assassin's Creed Heroine Is a Great Black Game Character. Here's How It Happened.,' *Kotaku*, 27 Feb. Available at: https://kotaku.com/this-assassin-s-creed-heroine-is-a-great-Black-game-cha-5987083 (Accessed: 27 May 2022).

Neely, E. (2017) 'No Player is Ideal: Why Video Game Designers Cannot Ethically Ignore Players' Real-World Identities.' *ACM Computers and Society* 47(3), pp. 98-111. https://doi.org/10.1145/3144592.3144602 (Accessed: 27 May 2022).

Patane, I. Lelgouarch, A. Banakou, D. Verdelet, G. Desoche, C. Koun, E. Salemme, R. Slater, M. Farnè, A. (2020) 'Exploring the Effect of Cooperation in Reducing Implicit Racial Bias and Its Relationship with Dispositional Empathy and Political Attitudes.' *Frontiers of Psychology*. Oct 28. https://doi.org/10.3389/fpsyg.2020.510787 (Accessed: 27 May 2022).

Patterson, J. (2020) 'Lessons from Time's ground-breaking two year immersive VR project, The March' FIPP, 30 March. Available at: https://www.fipp.com/news/lessons-from-time-vr-project-the-march/# (Accessed: 27 May 2022).

Ramos, D. (2020) 'The Messy Truth: Van Jones, Brie Larson and Elijah Allan-Blitz Talk The Immersive Empathy Of Eye-Opening VR Series,' *Deadline*, 27 August. Available at: https://deadline.com/2020/08/the-messy-truth-of-vr-experience-van-jones-brie-larson-elijah-allan-blitz-representation-inclusion-diversity-1203022976/ (Accessed: 27 May 2022).

Rathje, S. Hackel, L. and Zaki, J. (2021) 'Attending live theatre improves empathy, changes attitudes, and leads to pro-social behaviour.' *Journal of Experimental Social Psychology*, Vol95, July 2021, 104138. https://doi.org/10.1016/j.jesp.2021.104138 (Accessed: 27 May 2022).

Rueda, J. and Lara, F. (2020) 'VR and Empathy Enhancement: Ethical Aspects.' *Frontiers of Robotics and AI*, 7:506984. https://doi.org/10.3389/frobt.2020.506984 (Accessed: 27 May 2022).

Santos, D. (2020) 'Why the gaming industry is now one of the most exciting industries to work in,' Medium, 07 July. Available at: https://medium.com/@dsantos.rgf/why-the-gaming-industry-is-now-one-of-the-most-exciting-industries-to-work-for-rgf-fb8ca9aae1be (Accessed: 27 May 2022)

Smith, P. (2019) 'Into Iñárritu: How CARNE y ARENA Set the Bar for How VR Should Be Experienced,' *Immerse*, 08 August. Available at: https://immerse.news/into-i%C3%B1%C3%A1rritu-how-carne-y-arena-sets-the-bar-for-how-vr-should-be-experienced-de852124a6ff (Accessed: 27 May 2022).

Srauy, S. (2019) 'Professional Norms and Race in the North American Video Game Industry,' *Games and Culture*. 2019; 14(5): pp. 478-497. https://doi.org/10.1177/1555412017708936 (Accessed: 27 May 2022).

Stearns, E. (2019) 'Setting the Tone - Ethan Stearns, MWM Immersive (Carne y Arena, Chained),' Immersive Design Summit 2004, September. Available at: https://www.youtube.com/watch?v=RlvcUK8Iaew (Accessed: 27 May 2022).

Stenros, J. (2012) 'In Defence of a Magic Circle: The Social and Mental Boundaries of Play.' Proceedings of DiGRA Nordic 2012 Conference: Local and Global - Games in Culture and Society, Vol. 10, ISBN / ISSN 2342-9666.

Sunderland, P. (2019). *The Virtual Worlds of Cinema: Visual Effects, Simulation, and the Aesthetics of Cinematic Immersion*. PhD Thesis. University of Sydney. Available at: https://ses.library.usyd.edu.au/handle/2123/21052 (Accessed: 27/May 2022).

Tapley, K. (2017) 'Oscars: Alejandro G. Inarritu's VR Installation 'Carne y Arena' to Receive Special Award.' *Variety Magazine*, Oct 27. Available at: https://variety.com/2017/film/awards/oscars-alejandro-g-inarritus-virtual-reality-installation-carne-y-arena-to-receive-special-award-1202601265/ (Accessed: 27 May 2022).

Thériault, R. Olson, J. Krol, S. Raz, A. (2021) 'Body swapping with a Black person boosts empathy: Using VR to embody another,' *Quarterly Journal of Experimental Psychology*. https://doi.org/10.1177/17470218211024826 (Accessed: 27 May 2022).

Utichi, J. (2017) 'The Birth of An Art Form: How Alejandro G. Iñárritu And Emmanuel Lubezki Learned To Master VR – Cannes,' *Deadline*, 21 May. Available at: https://deadline.com/2017/05/alejandro-g-inarritu-emmanuel

-lubezki-carne-y-arena-virtual-reality-cannes-1202099184/ (Accessed: 27 May 2022).

Wiederhold, B. (2020) 'Embodiment Empowers Empathy in VR.' *Cyberpsychology, Behavior, and Social Networking*, Volume 23, Number 11, 2020. https://doi.org/10.1089/cyber.2020.29199.editorial (Accessed: 27 May 2022).

Chapter 3

An embodied poetics: Screenwriting for immersive media

Rafael Leal

Pontifical Catholic University of Rio de Janeiro, Brazil

Abstract

This chapter examines the process of screenwriting for VR environments. The author has 15 years of experience as a screenwriter and has worked on multiple VR projects. The chapter reports upon those experiences while arguing that VR requires screenwriters to think more consciously about the body of the spectator. Ultimately, the chapter analyses how the concept of the *embodied script* is developed and situated in the context of film production and questions the methodological separation between creation and execution found in typical approaches to screenwriting.

Keywords: screenwriting, embodied script, film, embodiment

This study was supported by the DAAD — Deutscher Akademischer Austauschdienst's Programme "Research Grants – One-Year Grants for Doctoral Candidates 20/21."

***

Introduction

In the time of the caves, our ancestors sat around the fire at night and told themselves stories, starting an ancient and noble tradition. This well-known cliché is repeated to exhaustion throughout screenwriting learning environments. The example is typically either used to reinforce the alleged primacy of storytelling in contrast to other narrative forms or to illustrate how storytelling is natural to mankind (among other similar questionable propositions). In an informal conversation about narration, the VR creator Oliver Czeslik made an outstanding remark about the cave myth: as it was probably very cold in there, they would likely be dancing around the fire to warm up before telling any stories.

I don't want to oppose dancing and story cognition, at least not for now, but Czeslik's idea helps us understand that storytelling is neither a natural nor necessary destiny for any medium or society. This is true for film as an art form, as postulated by Elsaesser (2018), for whom the move towards narrative feature films was made "gradually and for complex reasons: social (to attract a middle-class audience), economic (to raise the ticket value) and institutional-ideological (to transfer power from the exhibitors to the producers" (p. 174). And this is also true for VR and other emerging immersive and interactive media (IIM), in which the bodies of the creators and interactors play an unprecedentedly important role. I use the terms spectator when referring to the individual person in fruition of flat linear media, such as film and television, and interactor when applied to immersive and interactive media.

After 10 years of writing professionally for Brazilian film and television, in the last few years I have been developing a growing interest in creating for IIM, exploring this new media's boundaries, and perhaps contributing to their expansion. However, as the creative process for IIM is essentially different from traditional media, it exposes the limitations of a long tradition of screenwriting when applied to IIM. Since 2017, in addition to several corporate films in VR, I have written a series of fictional short film scripts, such as *Suburban Silence* (2017), *Victoria* (2018) and *Man with a Bag* (2019), as an attempt to understand and overcome these limitations. As part of this effort, this study aims to examine the role of the bodies involved in the poetic process of IIM, such as VR films through a phenomenological approach. This approach includes taking creation and *fruition* as radically embodied phenomena to propose what should be a screenwriting poetics for these embodied media.

The poetics of any medium is defined by David Bordwell (1989) as "the study of the finished work as the result of a process of construction" (p. 371). Considering screenwriting as the object of analysis, in Ian W. Macdonald's (2013) words, "the poetics is not about how to write screenplays; but about understanding the actual practices of how they are written, and the institutions, individuals and beliefs that lie behind them" (p. 2).

Throughout the 20th century, the focus on the story led to the formation of a traditional poetics centred on the specific internal coherence of the screenwork, something allegedly achievable only through the 'correct' use of the dramatic structure. Macdonald (2013) calls this the *Orthodox Poetics of Screenwriting*, "a set of certain assumptions about what the film narrative should look like, irrespective of content." (p. 37). It is conveyed mostly through manuals that are used in screenwriting courses all over the world: "manuals direct writers towards the 'right' way to do things. [...] Writers are expected to absorb the wisdom on offer, rather than question its basis" (ibid.).

One of the most famous and representative examples of this US-created prescriptive poetics is the international best-seller *Story* written by the Screenwriting guru Robert McKee (1997), who affirms in the introduction that, "*Story* is about eternal, universal forms" (p. 3). In other words, this kind of poetics evokes the authority of Aristotle's *Poetics,* uses successful movies as examples, and assumes that if a writer reproduces its principles, they will be able to produce 'correct' scripts. This makes it possible to extend David Bordwell's (1989) statement that the poetics of literature "has long been the province of sterile taxonomies and dogmatic prescriptions" (p. 387) to screenwriting.

But herein lies the problem: if the traditional screenwriting poetics, following its structuralist theoretical framework, focuses exclusively on internal elements of the screenwork, how can it possibly inform the creations in media that go beyond the limits of said framework and encompass external elements such as the interactor's body?

The script, used in film industries all over the world with minor regional variations, is—in its standard form—a document that describes the sounds and images that are offered as stimuli to the spectator. The only bodies that belong to the script, therefore, are those depicted on the screen; the spectator, therefore, is completely absent. *Fruition culture* is unimportant to orthodox screenwriting poetics, as the film (not to mention the script) is not impacted by the spectator's location, whether they are in a comfortable chair facing a big screen or standing up in the subway watching on their mobile phone.

But when it comes to IIM, the creative process drastically changes, as does the *fruition.* A VR film is a radically embodied experience, as its final form depends entirely on the interactor's body and how it moves in its *spectatorial choreography.* To understand the specificity and the requirements of an embodied poetics, this chapter approaches the act of creating immersive and interactive narratives in comparison to the act of experiencing them. The chapter also examines the poetic process of linear and flat narratives through the lenses of phenomenology.

According to Csordas (1990), phenomenology is "a descriptive science of existential beginnings, not of already constituted cultural products" (p. 9). It follows Merleau-Ponty's suggestion that the starting point of the analysis should be the individual (embodied) experience of perception, considering all its richness and indeterminacy.

This chapter aims to understand this new poetic system through a phenomenological approach to the material and technological experience of writing, which involves the writer's body as a means to explore and experience the work-in-progress as closely as possible to the way it will be experienced by the interactor, bringing awareness to "the dynamic and essentially correlational structure of that experience as it entails the existential activity of writing, the intentional objects that emerge in material form as the means of writing and the written matter, and an embodied and enworlded subject who is the writer" (Sobchack, 2004, p. 119).

At the same time, foreseeing how this poetics will allow the creation of a script where "activity, object, and subject are all enabled and mediated through a particular writing technology that spatially and temporally qualifies the embodied manner and objective style in which we write" (ibid.), this chapter proposes a new research paradigm for the field of IIM, which focuses on "a consistent methodological perspective that encourages reanalyses of existing data and suggests new questions for empirical research" (Csordas, 1990, p. 5).

Drawing on my own experience as a writer, and as a screenwriting researcher, I take into consideration three main fundamental elements in this analysis:

1. the bodies involved, which means not only the creator's and the interactor's body, but also the bodies depicted in the diegetic world;
2. the forces acting on these bodies, mainly meaning the tension between game experience and narrative fruition;
3. time and temporality, as in the different ways these bodies experience time throughout the poetic process.

Choreographing ghosts

Throughout my screenwriting career, I have never seen the spectator's body, not to mention the screenwriter's, considered in any screenwriting manual. This illustrates how the traditional screenwriting poetics approaches creation and *fruition* as purely mental processes, disconnected from bodies. In addition, writers involved in conceptualizing an artwork intended for the flat screen do not take into consideration their own bodies as part of the creative process, which is understood to be an exclusively mental activity. Nowadays, writing a script undeniably involves the corporeal act of inputting words into the computer through a human-computer interface (HCI) device such as a keyboard. Borrowing from my own experience, for example, it is not common for screenwriters to experience, or rehearse, their scenes before writing them, as a strategy to perceive them spatially, part of the creative process. Csordas (1990) argues "that on the level of perception it is not legitimate to distinguish mind and body. Starting from perception, however, it then becomes relevant (and possible) to ask how our bodies may become objectified through processes of reflection" (p. 36).

The screenwriter imagines the diegetic space as it should appear on the screen and describes it using written words. Nevertheless, the materiality of the writing experience is undeniable, as Vivian Sobchack (2004) notes when pain and other bodily feelings call her back from the computer "into a more physical world of writing and writing instruments, [and] reminds me also that my incorporation of pencil and pen and their particular materiality gives rise not only to particular spatial forms but also to particular temporal forms" (p. 114).

At the same time, screenwriters are not required to consider the bodies of their future spectators when writing a flat film: traditionally, they have never been required to do it, and their set of theoretical and practical tools has been proving itself more than enough for getting scripts written. In flat media, the exhibition space and the diegetic space are clearly distinct, separated by the screen: on one side, there is the spectator, sitting in a theatre or laying down on a couch at home, facing the screen; on the screen, projecting itself beyond the reflective surface, there is the diegetic space, a bidimensional representation (the screenwork) of a tri-dimensional scene where the story takes place. The spectators are not part of the diegetic space whatsoever and cannot navigate through it, their experience is limited to the stimuli defined by the creators: images conveyed within the frame and sounds coming from the speakers, following an exact precise order and rhythm.

When a screenwriter engages in writing a flat film, they can assume that the spectators will look at the diegetic world on the screen as if they were looking through a one-way window. From where they are, they cannot be seen by the characters inhabiting the diegetic space. But some films purposely break the screen barrier, often called the *fourth wall* and depict characters looking directly into the camera, giving the impression that they are effectively looking at the spectator. It can also be assumed that in most cases, the spectators do not have any agency and do not question their involvement in the story. Agency defines the interactor's capacity to act, and not only their ability to interfere in the narrative itself. It is a concept very useful in differentiating active reception, in which the spectator creates meaning from the affections and sensations produced by the screenwork's stimuli, from interactivity, regarding their ability to spatial exploration, media objects manipulation and nodal navigation. Unlike interactive media, traditional screenwork does not allow the spectator to participate, or, as VR creator Céline Tricart (2017) puts it: "we know we are outsiders" (p. 97). The knowledge she mentions, which I call *fruition culture,* was developed in the first decades of the 20th century and has remained without significant changes to this day; consequently, fruition culture fits with a model of screenwriting poetics that disregards the embodied spectator as a meaningful element in the creative process.

A screenwriter creating for IIM, however, faces a completely different set of challenges, starting from the overlap between the exhibition and the diegetic spaces and the increased level of agency available to the interactor. Instead of facing the diegetic space through the screen frame, when the interactors put the head-mounted device on, they can feel themselves inside the scene, completely surrounded by the diegetic space. They may or may not be able to see themselves in the scene, and in some cases, may only be able to see part of an avatar—such as hands or guns—that can provide them with some level of embodiment.

Embodiment defines how individuals inhabit the world (Csordas 1990), meaning our individual capacity to be present as a whole, rather than the sum of distinct elements and isolated processes occurring simultaneously. Regarding screenwriting for IIM, which consists in part in the creation of a diegetic world, embodiment refers to the ability to perceive oneself as a carnal body and makes explicit "how the very nature of our embodied existence 'in the flesh' lays the concrete foundations for a materialist — rather than idealist — understanding of Aesthetics and Ethics" (Sobchack, 2004, p. 3). And of poetics, I must add.

Although the interactor's body is carnally located in the exhibition space, they may be able to trick their sensory apparatus into perceiving themselves as being present in the diegetic space. This phenomenon is called *telepresence,* and as described by Brenda Laurel (1993), it "allows you to take your body with you into some other environment... you get to take some subset of your senses with you into another environment. And that environment may be a computer-generated environment, it may be a camera-originated environment, or it may be a combination of the two" (p. 162).

The exact level of embodiment will depend on many factors, especially those related to the technology used to create and experience the artwork, such as

- the precision of the interactor's relative position in real-time processing; the extent of the virtual body given to the interactor;
- the degree of realism provoked by the surrounding images;
- the amount of sensorial (including haptic) feedback offered to the interactor;
- and finally, the level of agency available to the interactor, which is reflected not only in the diegetic world's openness to manipulation and transformation but also in the media objects' (such as characters) capacity to acknowledge and react to the interactor's presence in the diegetic world.

Following Lucia Santaella's (2003) lead, I also avoid making a distinction between a *real* and a *virtual* body, "because there is no epistemological opposition more mistaken than virtual *vs* real, or virtual *vs* physical, as if virtual representations were not real and physical. The difference does not lay in being real or unreal, but in distinct types of reality and physicality" (p. 304). She proposes distinguishing between carnal bodies, such as the interactor, and alternative bodies, such as the characters, whose bodies are representations native to the digital diegetic world, including the interactor's avatar itself. I prefer to call them either carnal or diegetic bodies, as *diegetic* reveals more of their nature and origin than *alternative.*

Santaella (ibid.) echoes the concern of cyberspace theorists who "refer to this opposition [between the real and the virtual] as being the representational dilemma of the cyberspace, whose most intriguing aspect certainly concerns the boundaries of the body in the process of embodiment and de-embodiment that take place in VR, a dilemma that challenges not only Psychoanalysis but also Semiotics" (p. 304). As the interactor is freed from the limits of the frame and becomes able to look (and sometimes move) anywhere they want within the diegetic space, their *spectatorial choreography* directly interferes in the final form of the artwork, because it ultimately determines to which stimuli they will be exposed.

Moreover, the point of view (POV) has an impact on the agency and must also be taken into consideration as it deeply affects the interactor's experience. From what vantage point does the interactor observe the diegetic world, and what does this position mean relative to the context and to the story? Do they take the place of a character, such as the protagonist (first-person POV), or the sidekick (second-person POV)? Or are they a disembodied third-person POV, an invisible witness to the scene?

As compared to the orthodox poetics of screenwriting, in addition to losing the power of deciding what will be inside the frame, IIM creators also lose the power to impose a cohesive plot. An orthodox screenwork is driven by causality, which links the internal narratological elements and shapes the story as a cohesive linear stream of cause and consequence, whereas in embodied poetics, the story advances through the embodied actions of the interactor, who manipulates media objects. This can be either performing a gesture, activating a hyperlink, or clicking a button, making each *fruition* a unique experience.

Creating for IIM requires embodied thinking from the screenwriter, involving their own body in the creative process, testing their propositions and experimenting during the development process: "as we incorporate writing technologies, we simultaneously excorporate and give material form to our thoughts and feelings; and, as there is spatial reciprocity between the subjective and objective poles of this process, so also is there temporal reciprocity" (Sobchack, 2004, p. 115).

Time and temporality

In the aforementioned informal conversation with the VR creator Oliver Czeslik, I asked him about the duration of *Mind the Brain* (2021), which is his most recent VR creation. "Well," he said, after hesitating for a moment, "it depends a lot on who is experimenting it. I would say 15 to 40 minutes." It is very unlikely that a screenwriter working on a flat linear feature film would have similar doubts, as scripts usually have between 90 and 120 pages and there is a general belief — although not necessarily always true — that one script page will be equivalent to one minute of produced screenwork. Kathryn Millard brings up the discussion about film formatting conventions, referring to a specific aspect of the orthodox poetics of screenwriting, one that regards the way a script should look, in order to make its *fruition* time as close as possible to the final screenwork duration. It involves the use of the Courier New font in a specific pitch and size, giving the scripts a characteristic typewriter-style look that screenwriting software has been reproducing and even reinforcing. According to Millard (2010), "conventional wisdom in the film and television industries suggests that the screenplay is not only a creative document, but also one that encompasses production planning; providing information about locations, actors, sets, props, time of day and, most vital of all, timing. If the usual film formatting conventions are followed, then a page of screenplay equals one minute of screen time" (p. 19).

In flat movies, a common strategy to ensure this equivalence is a timed uninterrupted reading of the entire script, considering the rhythmic proposal made by the director, who paces the reading according to their vision. This process usually takes place in the beginning of the film's pre-production when changes to the script are made and the screenwork starts to be planned in a more concrete way. The process allows creators to foresee the duration of the completed screenwork's intended fruition. This scenario has been radically transformed with the introduction of computers, such as smartphones and TVs, into the spectatorial experience. Until a few years ago, new telenovela chapters were aired daily in Brazil, gathering crowds in front of televisions at a specific time. If someone missed it, it was lost, and the only solution was to wait for the following day's chapter. Conversely, with video-on-demand technologies, spectators take control of how and when they watch it, opening space to alternative *fruition cultures* such as binge-watching.

In a ground-breaking study about the popular Brazilian telenovela, screenwriter and researcher Rosane Svartman (2019) states that "we can consider three different temporal flows for all narratives: the story time – how time passes within the story; the discourse time – the duration and structure within a narrative; and the narration time – the time that the screenwork is being watched, the stipulated period to tell the story" (p. 220).

While watching a flat film, or a telenovela, regardless of where the fruition takes place (i.e., a theatre or a living room), the spectator experiments the narration time regardless of the story or discourse time. In Svartman's (ibid.) words, "until recently, the timing of the flow of a telenovela, be it the story time, discourse time or narration time, was not something the spectator could control" (p. 221).

This is directly opposed to what happens in IIM, where the interactor's embodied behaviour plays a central role in how the experience will unfold in terms of perceived time, as the conversation with Czeslik makes clear.

The spectator follows the events that happen to one or more characters in a predetermined pace and order, which means that they inhabit the present (narration time) and have no agency over the story or discourse times. Conversely, in IIM narratives, the interactor is bodily present both in the exhibition and the diegetic spaces, thus equalizing the story and discourse times. In this case, the interactor is largely responsible for defining the experience, as they can be endowed with enough agency to determine, or at least influence, the duration of the experience.

Forces and tensions

Since the beginning of the new wave of IIM in the last decade, artists and researchers have been dealing with an apparent contradiction between game experience and narrative contemplation. This is often referred to as the *ludonarrative dissonance,* a term coined by the game designer Clint Hocking (2007). In his words, it refers to "the leveraging of the game's narrative structure against its ludic structure all but destroys the player's ability to feel connected to either."

Mentioning a study by Katy Newton and Karin Soukup, filmmaker Celine Tricart (2017) notes that "when surrounded by a 360º sphere of potential information, participants [interactors] are less likely to catch the subtleties of the story, especially when those subtleties are delivered through audio" (p. 92). In the same direction, Mads Larsen's (2018) research shows that "whether the story is told in first or third person, there is a substantial trade-off between immersion and our ability to absorb plot and information" (p. 75).

But this phenomenon has been known for a longer time: Kath Dooley (2017) informs that it was described by Meyer in 1995, when "he rightly predicted that the VR form would create tension between interactivity and dramatic structure, as audience immersion allowed the audience to enter into a fictional world as either witness or protagonist" (p. 165). She highlights Meyer's suggestion that preserving the dramatic structure would require constraining the interactivity, as if opposing dancing and listening to stories.

I respect that the *spectatorial choreography* in IIM is not exactly a dance, but it is definitely closer to a dance than to watching a flat film in a cinema or on a couch. Now that the physical body projects itself, in an embodied manner, into the diegetic space, it becomes clear that the embodied presence is a complex stimulating process, one that requires some degree of attention that challenges the traditional narrative cognition. Try listening to the audiobook of *Ulysses* while dancing the tango.

I am not endorsing any binary opposition between critical distance (narrative cognition) and an embodied agency (immersion or telepresence), although some authors approach these two poles as mutually exclusive alternatives. One of them is Lev Manovich (2001), who proposes that "the oscillation between illusionary segments and interactive segments forces the user to *switch* between different mental sets — different kinds of cognitive activity" (p. 210). On the other side, I tend to agree with Maurice Merleau-Ponty (1962), who sees our embodied experience inseparable from the production of meaning and significance. In his words, "consciousness projects itself into a physical world and has a body, as it projects itself into a cultural world and has its habits: because it cannot be consciousness without playing upon significances given either in the absolute past of nature or in its own personal past, and because any form of lived experience tends toward a certain generality whether that of our habits or that of our bodily functions" (p. 137).

While Merleau-Ponty is talking about the embodied experience in the world, he makes evident the distinction between these two regimes, and suggests that they may happen concomitantly. For screenwriters creating for IIM, the balance between the critical distance — or cultural narrative cognition — and the embodied presence is not only a concern: it is the core of the experience design. This is the reason why a traditional script does not fit the creative process for IIM: it does not allow the screenwriters to engage their own bodies in the projected experience.

The embodied script and film production

The traditional production process of creating a flat film is commonly divided into three distinct well-known steps: pre-production, which means planning and preparation; production itself, which refers to the shooting phase; and finally, post-production, where the footage is assembled into a screenwork and delivered as a product. In this scheme, when the film production starts, the script is already written. One can argue that the development and funding steps take place before production, but regardless, this three-step model reflects the common belief that the screenwriters' office does not belong to the film production headquarters: they are, at best, neighbours, or the previous tenants.

However, considering the increased digitalisation of the film production process, the need for distinct isolated steps surrenders to the advantages of a simultaneous, recursive approach to production that integrates teams in a more dynamic way and allows new ways of creative collaboration and problem-solving. This is particularly important when noting the growing use of game development technology in film production. To date, the main example is *The Mandalorian* (2019), shot using a game engine to create the virtual scenarios, offering the actors the ability to manipulate digital objects and effects processed in real-time.

From the examples I have seen in my career, an IIM production process is much more similar to software development than to a traditional flat film production. One of the main similarities is the way teams are distributed in a less hierarchical structure, with multidirectional communication and progress-tracking through project management tools. As the teams' works start to be assembled, there are always new issues coming up, and some of them might require involving more than one team in the solution, while others may require changes in the original project itself. In IIM production, these decisions may interfere with the story or with the interactor's experience, and this produces a series of come-and-goes between script development and technical implementation. It allows for sharing the creative decision within a wider set of artists and discussing how to attract and direct the interactor's attention, which Dooley (2017) considers "perhaps one of the most important points for consideration by the writer and/or director" (p.168).

And being such an important point of consideration, it is better that it is examined by a multidisciplinary group of specialists instead of a sole author. IIM production involves intense back-and-forths between screenwriters and other professionals, as well as iterative testing and prototyping. Conversely, in a traditional film, the screenwriters' involvement often ends before pre-production and excludes them from the unfolding of the creative process. However, new technologies can often shift these current traditional processes and force teams to integrate the story specialists into the whole screenwork development. Or, as Millard (2010) points out, "pre-production, production and post-production tend to get collapsed into a single more fluid stage, in which images and sounds can be reworked to a much greater degree. [...] Surely then, more than ever, the screenplay needs to be a flexible document" (p. 16)?

Before addressing this important question, it is important to note that I am observing only the production process aspects, which deserve further study, and purposely ignoring all artistic and authorial aspects, which are much more important in the film business than in the software industry. In fact, as Kath Dooley (2017) remarks, artists responsible for creating and developing IIM narratives, especially those coming from areas different than traditional filmmaking, "may not use this term [screenwriter] themselves, preferring instead 'developer,' 'director,' 'content creator' or 'narrative designer,' the latter term being common in the gaming industry" (p. 163).

Even defining what a script is can be a complex task that exceeds the scope of this chapter. Furthermore, as I have taken a phenomenological approach here, this study is not concerned with seeking essences or hard definitions, but with a questioning of "the meaning of experience as it is embodied and lived in context — meaning and value emerging in the synthesis of the experience's subjective and objective aspects" (Sobchack, 2004, p.2). Therefore, there are two binomial script functions that may be useful to understand the new roles of the embodied script in film production: *architecture* and *run* (Bode and Dietrich, 2013).

Despite authors such as Millard (2010) and Maras (2009) having criticized the worn-out metaphor of the script as a blueprint, the related term *architecture* may be especially useful in IIM studies as it describes, with a spatial hint, the map of navigational possibilities. It is undeniable that the discipline of architecture encompasses much more than any blueprint, which cannot describe the experience of a body navigating through the projected building. It is not a coincidence that IIM have been widely (and increasingly) used in the real estate business to showcase the property to potential buyers in a very realistic way, allowing convincing embodied visitation for many different places without having to be physically present there.

When it comes to screenwriting, architecture means the objective mapping of the nodal structure as well as the projected diegetic spaces and the ability to navigate through and between them, and the run refers to the subjective and temporal experience of *fruition*. In a linear flat film script, the architecture is very similar to the run. As you read it, you imagine the experience: it is often said in screenwriting that the script should provoke mental images of what will be depicted on the screen. In IIM, nonetheless, architecture and run mean pretty different things. While the architecture encompasses all the navigational possibilities, each run describes one unique path among the many possibilities, choice by choice. In Bode and Dietrich's words (2013), it is a "transformation from potentiality to actuality," where the potential choices are "dissolved into a particular event, a new situation" (p. 67).

As a practitioner, I am always interested in the processes that will lead my creations to their production as screenwork, in addition to the theoretical approaches to this subject. Here I will mention one last - but not less important - difference between the traditional and the embodied poetics of screenwriting. Before becoming a set of instructions for a cast and crew, a script is a marketing piece, directly involved in the funding process that takes place prior to production. In a still earlier stage of development, the script frequently conveys the screen idea through a description of scenes (Macdonald, 2013). The potential funding executives are used to read scripts submitted in this format and are (hopefully) able to grasp the proposed narrative in general terms and make a decision regarding this production.

In IIM, however, the funding mechanisms are still not as established as they are for the film industry, and an embodied script can be much more difficult to assess. This happens not only because the script lacks a conventional traditional form to which stakeholders are accustomed, but mainly because its architecture contains a multiple number of possibilities that lead to potentially distinct outcomes. And here is where the *run* comes: because it refers to a specific *fruition* episode and embodied visit to the diegetic space, it offers an account of the experience that, although limited to one among many possibilities, gives the reader the chance to imagine how the interactor might feel while experimenting this projected immersive and interactive screenwork.

Conclusion

Until recently, screenwriters had seldom been required to think of the creative process as an embodied phenomenon. The embodied poetics for IIM brings to script development a new set of goals and concerns, such as the goal of producing embodiment and building presence. They allow the interactors to experience the diegetic world, as if they are effectively part of it, from a specific point of view. Transported to the diegetic world, the interactors do not relate to the story only through empathy towards some character, but they also experience through their own body the engaging potential of the emerging IIM. Consequently, this new screenwriting poetics incorporates phenomenological foundations and allows understanding of creation and reception also - and mainly - in their embodied, incarnated dimension.

If traditional poetics sees the script as a static structure, like a mosaic whose pieces are meticulously arranged by the artist to encode an image, a linguistic approach to the new poetics will see the script as a dynamic system. In this system, its constituent elements are related through a complex, non-hierarchical process, like a mobile — a kind of kinetic sculpture that, when manipulated, shows itself in a myriad of ways, depending on the forces applied by the person manipulating it and the point of view of the interactor. Unlike the mosaic, the mobile assumes distinct forms through interaction.

Accordingly, the embodied poetics can be compared to the device known as a flipbook: a booklet whose pages contain sequential, progressive and slightly different images; an ingenuity that, as occurs in a movie projector, exploits the retinal persistence, a physiological characteristic of the ocular apparatus, to give the illusion of movement. When flipped through, the flipbook images come to life — a life that is not contained merely in its pages, but in its embodied interaction.

Perhaps exceedingly allegorical when compared to the IIM script development process, the mobile and the flipbook images reinforce the importance of the body in the *fruition*. The embodied approach to creation and *fruition* is the core of a poetics that entails the need to consider the interactor's body from the beginning of the script development. This creative process is not restricted to the screenwriter's mental activities and involves their own body. It then becomes a new way of approaching creation and production for IIM.

IIM has been around for some time now, but the technological developments that allowed this recent wave of creativity are still at an early stage of evolution. As happened with flat films a little over a century ago, this medium's capacity to expand our consciousness and the way we tell and experiment stories is underexplored. Notwithstanding, in the prophetic words of Manovich (2001), the exploration of IIM experiences entails the desire for "a return to the primitive happy age of pre-language, pre-misunderstanding. Locked in VR caves, with language taken away, we will communicate through gestures, body movements, and grimaces, like our primitive ancestors" (p. 58).

Almost as if we were dancing around a fire.

References

Bode, C. and Dietrich, R. (2013) *Future Narratives: Theory, Poetics, and Media-historical Moment.* Berlin/New York: De Gruyter.

Bordwell, D. (1989) Historical Poetics of Cinema. *The Cinematic text: methods and approaches (Georgia State Literary Studies).* 3. New York: AMS.

Csordas, T. J. (1990) Embodiment as a Paradigm for Anthropology. *Ethos.* 8(1) pp. 5-47.

Dooley, K. (2017) Storytelling with VR in 360-degrees: A new screen grammar. *Studies in Australasian Cinema.* 11(3). pp. 161–71.

Elsaesser, T. (2018) *Cinema como arqueologia das mídias.* São Paulo: SESC Editions.

Hocking, C. (2007) *Ludonarrative Dissonance in Bioshock.* Available from: https://clicknothing.typepad.com/click_nothing/2007/10/ludonarrative-d.html (Accessed: 10 January 2021)

Laurel, B. and Coyle, R. (1993) *The Genesis of VR,*" in *Future Visions: New Technologies of the Screen,* edited by Philip Hayward and Tana Wollen. London: British Film Institute.

Leal, R. (2020) Poética do Roteiro Audiovisual: uma abordagem fenomenológica para narrativas imersivas. Revista Moventes. 5. Available from: https://revistamoventes.com/2020/05/31/poetica-do-roteiro-audiovisual-uma-abordagem-fenomenologica-para-narrativas-imersivas/

Macdonald, IW. (2013) *Screenwriting Poetics and the Screen Idea.* London: Palgrave Macmillan.

Manovich, L. (2001) *The Language of New Media.* Cambridge: MIT Press.

Maras, S. (2009) *Screenwriting — History, Theory and Practice.* London: Wallflower Press.

McKee, R. (1997). *Story: substance, structure, style and the principles of screenwriting.* New York: HarperCollins.

Merleau-Ponty, M. (1962). *Phenomenology of perception.* London: Routledge.

Millard, K. (2010) After the typewriter: the screenplay in a digital era. *Journal of Screenwriting.* 1(1). pp.11-25. Available from: http://dx.doi.org/10.1386/josc.1.1.11/1

Santaella, Lucia. (2003) *Cultura e Artes do Pós-Humano: da Cultura das Mídias à Cibercultura.* São Paulo: Paulus.

Sobchack, V. (2004) *Carnal thoughts: embodiment and moving image culture.* Los Angeles: University of California Press.

Svartman, R. (2019) T*elevision in Transformation — How the telenovela can indicate strategies for corporate television in Brazil in the face of transformations in spectatorship, convergence of media and interactive platforms.* PhD Thesis. Universidade Federal Fluminense.

Tricart, C. (2017) *VR Filmmaking: Techniques & Best Practices for VR Filmmakers.* New York: Routledge.

Chapter 4
Taking stock: VR in journalism

Hannah Greber

University of Vienna, Austria

Abstract

Since the introduction of Virtual Reality (VR) to journalism, it has undergone a phase of experimentation. After a decade of learning, it is time to take stock and ask how VR journalism is produced, used, and received, and how this practice might contribute to the democratic role of journalism. Drawing on literature, I propose the following. First, the application of VR requires a reconsideration of traditional journalistic storytelling. Second, there is a lack of viable financing strategies. Third, the effects of VR in journalism are more nuanced than expected. Finally, core understandings of journalism are challenged by VR journalism.

Keywords: Virtual Reality, Journalism, Immersive Journalism, 360-degree videos

***

Taking stock: VR in journalism

VR journalism refers to a type of journalism that uses inclusive technology, such as VR and 360-degree videos (Watson, 2017), to create an all-encompassing experience of the news (Buji and Hamari, 2020a). In 2010, the journalist Nonny de la Peña and a team of colleagues published one of the first-ever applications of VR in journalism, named "*Gone Gitmo.*" In this short production, the audience adopts the perspective of a prisoner in the infamous Guantanamo Bay prison. Uniquely at this time, the use of VR to tell a journalistic story "afforded unprecedented access to the sights and sounds, and possibly, the feelings and emotions that accompany the news" (de la Peña et al., 2010, p. 292).

Importantly, VR journalism offers the possibility to interact with a story through immersive narratives that provide a more active role within the story, such as a first-person narrative (de Bruin et al., 2020). The inclusive nature of the VR technology also aims at bringing a sense of presence into the journalistic sphere; that is, to establish a sense of 'being there'—being at a scene in a news environment (de la Peña et al., 2010). Currently, three forms of journalistic VR stories have been identified. These include 1) 360-degree live stories, 2) 360-degree news, and 3) 360-degree documentaries (Sirkkunen et al., 2020). While 360-degree videos are not VR, in the context of journalism and if used with a head-mounted display, 360-degree videos have similar capabilities as VR videos. They are subsequently considered within the umbrella term "VR" (Davis and Millet, 2021).

Concerning the choices of topics, VR journalism often focuses on non-time sensitive topics (Vindenes and Gynnild, 2020), such as culture, environment, and criminality, but also war and immigration are covered frequently (de Bruin et al., 2020).

Shortly after the introduction of VR to journalism by de la Peña, there was great interest within the field of journalism to utilize VR videos to bring the news closer to the audience. Following a steady increase in sophistication of the technology involved in VR journalism (Watson, 2017, p. 8) and the assumed potential of VR journalism to function as an "empathy machine" (Milk, 2015) news outlets such as *The Guardian, The New York Times* and *Euronews* experimented with this technology, with the aim to provide the audience with an all-encompassing and engaging experience of the news (Bujić and Hamari, 2020a). Yet, after early enthusiasm in the field during 2017 (Sirkkunen et al., 2020), the use of VR in journalism has declined. To this end, in late 2019, the *New York Times* deactivated their Daily 360-degree application (Palmer, 2020), and VR news productions by *Euronews* declined from 73 produced videos in 2017 to 20 in 2019 (Sirkkunen et al., 2020).

At this juncture, and following a period of experimentation, it is time to evaluate what we now know about the use of VR in journalism. More specifically, I ask 1) how journalistic VR videos were produced, used, and received, and 2) how this form of journalism potentially contributes to the democratic role of journalism. To answer these questions, I discuss the narrative form before focusing on the production process of VR journalism. I then provide an overview of the effects and audience perspective of VR journalism. Subsequently, the question of the democratic role of VR journalism is addressed. Ultimately, I draw conclusions based on how VR influenced the field of journalism and formulate thoughts on next steps. Based on a review of existing studies on the use of VR or 360-degree videos in journalism, in this theoretical chapter, I argue, first, that the use of VR to tell journalistic stories has had a considerable influence on storytelling techniques; second, that the demise of interest in the production of VR stories is connected to their lack of financial viability; third, that the hype about the assumed effects of VR journalism has so far not been confirmed by empirical observation; and finally, that VR journalism challenges core notions of what journalists should and should not do as observers of society, thereby presenting a healthy development within a profession characterized by strong professional norms and routines.

VR journalism: The birth of a new journalistic genre?

VR as a medium and its application to journalism requires a reconsideration of how journalistic stories are told to the extent that it can be seen as a new narrative genre. Understanding this potential goes beyond applying VR to traditional journalistic stories, as the narrative needs to be transformed into an interactive experience (Domínguez, 2017). While the empowering role of agency was more prevalent in VR journalism of the early stages (e.g., de la Peña et al., 2010), limited forms of agency, such as "looking around," still require a reconsideration of narrative techniques. Including agency possibilities in the experience of news means that 1) the narrative of VR journalism becomes "place-based," 2) factual information has to be used scarcely, 3) new strategies to guide the attention of the audience are required, and 4) the first-person narrative perspective becomes more important.

First, news told with VR increasingly become slow-paced and "place-based" (Dowling, 2020, pp. 99). Due to the agency of the audience in VR, the surrounding space becomes more important. In VR journalism, the audience can look around and explore each scene. This exploratory behaviour is subsequently related to "lower levels of focused attention and poorer recognition and cued recall of information" (Barreda-Ángeles et al., 2020, p. 12). Consequently, journalists take this curiosity into account when producing VR journalism, and mainly choose to cover news with VR when the environment is extraordinary or important to the story (Sirkkunen et al., 2020), such as showing the aftermath of an earthquake (CNN, 2016) or historical artifacts (AP VR-360, 2016). The exploratory behaviour of the audience also means that editing techniques, such as quick cuts, cannot be used; thus, VR journalism is generally a slower-paced form of journalism, using fewer shots than video journalism (Vindenes and Gynnild, 2020).

Second, journalists found that they needed to limit the information in the form of hard facts for VR stories (Mabrook, 2021). In VR, the audience is ideally fully immersed in a story. This in turn means that due to bodily involvement, the audience processes the information differently than in the context of videos (Barreda-Ángeles et al., 2020). As a consequence, content creators reduce the so-called 'information overload' by only including a few selected facts in a VR story (Mabrook, 2021).

Thirdly, new strategies to guide the attention of the audience are required. Agency in the form of 'looking-around' is seen as being "not interactive in any significant sense" (Palmer, 2020, p. 15). More sophisticated approaches towards the implementation of agency and control, such as interacting with a character or choosing a storyline, could guide the attention of the audience. Through meaningful interaction, the audience would become part of the story rather than being a passive bystander (Wu et al., 2021).

Finally, the first-person narrative perspective plays an important role in journalistic VR, as this helps the audience to feel as if their own body was transported to the virtual environment (Sánchez Laws, 2019). The initial claim for the use of this form of journalism was that through the use of VR, the audience could experience the news as if it happened to them, thereby increasing engagement (Sánchez Laws, 2019) and empathy (Archer and Finger, 2018). Despite this, until now, the implementation of the first-person narrative perspective remains the exception rather than the norm, leaving the audience "stuck on the sidelines, only able to watch the story a journalist wants to tell, albeit in a 360-degree view" (de Bruin et al., 2020, p. 15).

To create VR journalism, narrative conventions and storytelling techniques need to be adapted to the logic of VR. Thus, the genre of VR journalism focuses on slow-paced shots of extraordinary locations, rather than densely packed, quickly cut videos. Thereby, it moves from telling a story to letting the audience experience a story (Maschio, 2017). However, these implementations require complex technological processing (Wu et al., 2021), which makes it difficult for journalists to create a truly immersive and interactive experience.

VR and the journalistic production process

Since the introduction of VR to journalism by de la Peña, a decade of slow adoption and industry-pushed experimenting with the technology has passed, yet VR journalism is far from being mainstream (Watson, 2017). Initially met with great interest, recently the number of news stories told with VR has declined (Sirkkunen et al., 2020). This development can be connected to three overarching issues: 1) a challenging and costly production, 2) lack of a viable financing scheme and, 3) the absence of aim for effects on the audience.

Challenging production cycles

The production of news in VR is challenging to journalists and their collaborators and can be described as "complicated and expensive" (Sirkkunen et al., 2020, p. 18). Despite the technological advances of the last decade (Watson, 2017), producing stories for VR poses several issues to journalists. While journalists undergo a training phase in some newsrooms (for instance, Euronews; Vindenes and Gynnild, 2020), many other journalists lack the knowledge and experience to produce news for VR, creating the need to collaborate with an external production company - but only a few do so. If news agencies collaborate with production companies, the quality in the form of more interaction possibilities and first-person perspective narratives improves (de Bruin et al., 2020). Yet, external collaborations possibly influence the editorial autonomy of journalists (Aronosn-Rath et al., 2016).

Studies on the workflow of journalistic VR highlight the cumbersome technology (Aronson-Rath et al., 2016), the lengthy production cycle due to the need for intensive preparations, long shootings, and time-intensive post-production. Additionally, the lack of editing possibilities are stressed (Harris and Taylor, 2021). Hence, the expensive, effortful and lengthy production stands in contrast to an industry that is typically known for its fast-paced, 24h news cycle (Deuze, 2005) and its economic constraints in a digital world (Newman et al., 2020). Due to all of this and the relatively nascent and fast-evolving field of VR journalism, many journalism agencies have chosen to not participate in the early experimentation with VR journalism (Watson, 2017).

Lack of viable financing scheme

With much time and effort put into producing immersive stories, ideally, these types of stories would pay off, economically speaking. However, no viable financing scheme to monetize VR journalism has emerged. In the early stages of experimentation, it was notably companies such as Samsung and Google who financed the production of journalistic VR stories, e.g., for *The New York Times* and *Euronews* (Sirkkunen et al., 2020, p. 13). Additionally, *The New York Times* delegated "the production cost to their corporate customers" (Ilvonen et al., 2020, p. 120) by mainly producing blended branded content. Beyond such funded experimentation, the monetisation of VR journalism remains challenging, as nobody "has yet cracked either ad-or subscription-based models for making the technology pay" (Watson, 2017).

In their analysis of value generation through VR, Ilvonen et al. (2020) highlight that effortful production requires a clear value creation, but they stress the evaluation needs to go beyond direct profits. For instance, VR journalism could provide a platform to attract audiences, which ideally turn into subscribers (Ilvonen et al., 2020). Nonetheless, the case of *The New York Times* shows how complex evaluating the economic value of VR journalism is. While one common argument to use VR for news is to increase audience engagement (Ilvonen et al., 2020), a study on the popularity of and engagement with VR videos and normal videos by the *New York Times* indicates the opposite to be true. 360-degree videos were not as popular as traditional videos, perhaps due to a smaller audience being familiar with the technology and able to engage with these media formats. Furthermore, the audience was less likely to comment, share, or like a 360-degree video in comparison to traditional videos (Wang, Gu, and Suh, 2020). Thus, it is questionable whether the use of 360-degree videos would translate to subscriptions.

Does VR have the promised effects?

The use of VR in journalism was carried by the idea that this technology could transport the audience to a different reality—potentially creating empathy and a deeper understanding of the news (de la Peña et al., 2010). Additionally, the possibility to autonomously explore a scene was thought to increase the credibility of this form of journalism, thereby addressing the trust crisis of journalism (Kang et al., 2019). The indispensable evaluation of whether the technology held up to its initial promises, however, reveals a more complex picture.

To reiterate, one of the reasons VR journalism is used is the assumption that it transports the audience to a scene (de la Peña et al., 2010). Indeed, virtual environments outperform less immersive media formats when it comes to the sense of presence (Sundar Kang, and Oprean, 2017; Kang et al., 2019) and the feeling of embodiment (Ahn et al., 2016). Additionally, embodying an avatar leads to more presence and subsequent engagement with a topic (Slater et al., 2018). The sense of presence, then, translates into several impacts. For instance, a stronger sense of presence means that the audience is more likely to confirm VR as a technology (Shin and Biocca, 2018), and follow up on the content experienced in a journalistic VR production (Steed et al., 2018). Additionally, a stronger sense of presence is related to a stronger sense of empathy (Shin, 2018), to more enjoyment and more subjective involvement with the VR experience (Van Damme et al., 2019). Overall, presence is an integral part of VR journalism that is triggered by immersive technology and can translate into beneficial effects.

Famously claimed to be the "empathy-machine" (Milk, 2015), the audience reports stronger experiences of empathy after a journalistic VR production in comparison to text and picture stimuli (Archer and Finger, 2018; Sundar Kang, and Oprean, 2017) and in comparison to less interactive stimuli (Schutte and Stilinović, 2017). Other findings indicate that there is no significant impact of the level of immersion on empathetic reaction (Herrera et al., 2018), but that empathy is rather predicted by the evaluation of importance by individuals (Steinfeld, 2020). Similarly, the positive effect of presence on empathy is ruled out by the sense of enjoyment, which is related to the experience of an immersive production (Barreda-Ángeles et al., 2020). Hence, whether VR journalism enhances the experience of empathy is unclear.

Research investigating the effect of viewing a VR story on memory and recall indicates a negative impact. Several studies report lower free and cued-recall (Sundar, Kang, and Oprean, 2017) and memory and understanding (Vettehen et al., 2019) when comparing inclusive to less inclusive VR stories. Concerning long-term influence, however, one study shows that immersive treatment conditions were remembered better than non-immersive treatments when individuals reported a higher empathetic response (Archer and Finger, 2018).

Concerning the impact on credibility in form of accuracy, factuality, and trust VR and 360-degree videos are evaluated as more credible (Kang et al., 2019; Vettehen et al., 2019) and trustworthy (Sundar et al., 2017) than less immersive formats.

In the main, studies on the impact of VR journalism on the audience suggest no clear-cut effects. While the use of these technologies might enhance empathy and presence for some, personality traits still strongly influence the reaction to any VR journalistic story (Shin, 2018). Nevertheless, if the feeling of presence is successfully triggered in the audience, stronger effects can be achieved.

The audience as litmus test for VR

While the effects of VR are of importance, one arguably more important aspect for VR journalism tends to be neglected: the wishes of the audience. Whether or not the audience appreciates, values and ultimately buys and uses VR journalism is key to the evaluation of any technology (Ilvonen et al., 2020). Despite this fact, very little research on characterizing the audience of VR journalism has been conducted (Watson, 2017). Nonetheless, a few exceptions can be mentioned here. When asked, the audience assesses VR journalism as trustworthy and appreciates the ability to experience - and interact with - a story (Nielsen and Sheets, 2019). However, they also report feelings of intruding the private space of the characters in a story, as well as concerns about having missed important information (Jones, 2017). Lastly, experiments have shown that the audience enjoys journalistic pieces with a higher immersiveness (Van Damme et al., 2019; Vettehen et al., 2019).

The expectations of the audience when being presented with VR devices seems to surpass the quality of the actual implementations of VR journalism (Harris and Taylor, 2021). Low-quality 360-degree videos might not meet audience expectations and even lead to physical discomfort, ultimately decreasing audience engagement ((Wang, Gu, and Suh, 2020). There is a need for high-quality, newsworthy and easily accessible VR journalism "to sufficiently impress digital audiences." (Kang et al., 2019, p. 309). Nevertheless, satisfaction with the technology seems to have a minor influence on the continuance intention, which remains stronger for 360-degree videos in contrast to 2D videos (Bujic and Hamari, 2020b).

Much work remains when trying to understand who the audience of VR journalism is, whether, how, and why they appreciate it. Answering these questions could help the future development of journalistic VR experiences. To further inform prospective productions, we also need to understand to which extent VR journalism fulfils the democratic role commonly ascribed to journalism.

How could VR contribute to the democratic role of journalism?

The question of what journalism is and what it should and should not do to fulfill its democratic role is an ongoing debate and depends on the different societal, historical, institutional, and technological contexts surrounding journalism (Hanitzsch, 2007). However, a few basic notions of what journalism should do have been identified: journalism should inform citizens concerning issues that are relevant to them, so that they can form an informed opinion and engage civically in a democracy (Ryfe, 2020). Journalism ideally does so by providing a public service; being objective, neutral and as a result, credible; being autonomous and independent; having a sense of actuality, immediacy, and speed; and having a sense of ethics (Deuze, 2005) such as reporting the truth, minimizing harm and being transparent and accountable (Society of Professional Journalists, 2014). Along with these notions, certain practices were established, for instance, the style of the inverted pyramid, focusing on the facts of a story, and presenting the news in a balanced way (Ryfe, 2020). Let me now apply this understanding of journalism to VR journalism.

Focusing on the production process, traditionally, journalists try to report just the facts in a balanced way, utilizing an inverted-pyramid format. VR journalism strays from this norm by putting the experience of the audience at the centre of attention. In VR journalism, the focus drifts from storytelling to storyliving (Maschio, 2017). This new screen grammar also means that the information transmitted in a journalistic VR story is mostly based on experience—for instance, the information about the location of a story is denoted from being surrounded by the location, rather than by being told where something happened. Hence, VR journalism has a complex structure of information transmission (Rodríguez-Fidalgo and Paíno-Ambrosio, 2020). Additionally, journalists try to limit the amount of factual information included in VR pieces (Mabrook, 2021), which further limits the notion of a just-the-facts-report of a story.

Some of the core notions of journalistic norms as described above are objectivity, neutrality, or accuracy (Deuze, 2005). These norms help journalists to validate the claim to report the truth (Aitamurto, 2015). Initially, the practice of objectivity was seen as a strategic ritual to increase the credibility and trustworthiness of journalism (Tuchman, 1972). While the question of objectivity is debated in journalism generally (Shoemaker and Reese, 2014), the introduction of VR to journalism intensifies this discussion. In this format, objectivity is threatened by two levels of subjectivity: journalistic subjectivity, and the subjectivity of the user (Mabrook, 2021). On the level of the journalistic subjectivity lie the editorial choices made by the journalists often involving extraordinary environments (Sirkunen et al., 2020) as well as editing and stitching processes. For the audience to be able to trust these subjective decisions of journalists, editorial choices should be made transparent (Kool, 2016). On the level of the user subjectivity, VR provides a personalized experience for each individual by allowing them to control their line of sight within the experience. This could have repercussions for instance when trying to convey complex stories, as individuals have different VR experiences and thereby might have different understandings of the same story (Mabrook, 2021). Overall, then, VR journalism is considered to be a highly subjective rather than an objective form of journalism (Mabrook, 2021).

In contrast to the expectations of objectivity to be related to greater credibility, the often subjective nature of VR journalism is perceived by the audience as more accurate and more credible than non-immersive formats (Kang et al., 2019). Aitamurto (2018) refers to this as the credibility paradox. On the one hand, VR journalism entails making choices about content, editing, staging, choices of scenes, and the positioning of the journalist. On the other hand, in the case of real-life footage, the audience perceives the VR story to be more accurate than non-VR content, as they can look around and are not bound to one predetermined frame. VR journalism in this respect can be set within the tradition of visual journalism, which claims to support journalism through an eye-witness account that shows the world "as it is," while they often carry multiple interpretations and meanings (Zelizer, 2012). The inclusive and interactive form of VR journalism adds a layer to the perception of accuracy. The audience perceives VR journalism to be more accurate, while this form of journalism does not necessarily present a more accurate account of the news. VR in journalism is used to "reinforce the illusions of objectivity and accuracy in visual journalism" (Aitamurto, 2018, p. 15).

VR journalism also expands considerations related to well-being, based on the notion that journalism should minimize harm. Issues such as motion sickness question the physical comfort of VR (Rebenitsch and Owen, 2016). Moreover, journalistic VR experiences could be psychologically distressing. Kool (2016) stresses that "through an empathy-engendering the machine, the viewer becomes emotionally vulnerable" (p. 7); yet when tested, VR journalism does not influence distant suffering (Van Damme et al., 2019). Whereas existing guidelines regulate to some extent how to manage distressing events in the news, the bodily involvement of VR journalism makes this a bigger issue (Sánchez Laws, 2019). Nonetheless, while VR is found to be an effective medium (Diemer et al., 2015), whether these effects translate beyond emotional states to the well-being of individuals remains unclear.

Lastly, due to the financing by big tech companies (Sirkkunen et al., 2020), and the need to collaborate within news agencies (Watson, 2017), autonomy and independence are important issues that need to be addressed. By including different actors in the production processes, the use of immersive technologies could influence editorial decision-making. While this issue remains under-researched, the discussion by Aronson-Rath et al. (2016) indicates a complex interchange of decision-making processes between editors and producers.

Conclusion and outlook

Few innovations within journalism have the potential to drastically change the way journalism is done and challenge the core norms of journalism. Since the advent of VR in journalism, the perceived potential of this technology for journalism has been very clear: to let the audience experience a news event as if they were there. This is arguably achieved through high-quality technology, ideally a photorealistic scene that the audience can interact with in a meaningful way. Here, audiences not only understand the facts about a story but also how it feels like to be in a situation. This hope was initially guided by the idea of the "empathy-machine" (Milk, 2015). Yet some years on, empirical studies about the actual implementation and effects of VR journalism have shown a different picture.

First, there is hesitancy in truly adapting the logic of VR productions into the production of news. Despite the perceived importance of embodiment and the first-person perspective of VR journalism, only very few journalistic productions have truly implemented first-person perspectives into their stories, and even fewer have provided the possibility of embodiment through including an avatar in a story (de Bruin et al., 2020). This would be one of the core features of VR journalism to impact presence and the feeling of connectedness (Ahn et al., 2016).

Second, the perceived positive effects of VR journalism have not been confirmed so far. Empirical studies show that the audience does not seem to recall facts of VR stories very well. Therefore, immersive journalism should ideally be supplemented with additional, textual information. Furthermore, the initially emphasized audience focus is largely missing, with most interactivity possibilities remaining on a basic level (Palmer, 2020). However, it would be meaningful interactivity that would empower the audience.

Third, the implementation of VR has transformed understandings of what journalism is and does. First, the implementation of VR in journalism has set off a new genre of journalism focusing on the experience of the audience: immersive journalism. Initially used to refer to journalistic stories created in VR to "gain first-person experiences of the events or situation described in news stories" (de la Peña et al., 2010, p. 291), this term now extends to this approach towards journalism that can be characterized not only by the level of inclusion, but also by immersive narratives and interactivity possibilities (de Bruin et al., 2020). This definition then extends to the use of a spectrum of inclusive technologies, such as Augmented Reality and immersive audio, and further highlights the importance of the narrative dimensions. This approach allows for a broader conceptualisation and application of immersive journalism, while keeping in mind the core notion of this type of journalism: to make the news an all-encompassing experience (Bujic and Hamari, 2020a).

Additionally, VR journalism has shifted the focus back to the audience. It showcases the need for journalism to not miss "the next big thing" (Watson, 2017) and highlights the need to focus on how news can be catered to the audience in a way that is valuable, intelligible, but also enjoyable to them. Reports remain inconclusive—while higher immersion might lead to more enjoyment of a story, the question of enjoyment might be inappropriate when dealing with issues of suffering (Van Damme et al., 2019). Relatedly, agency seems to be the key to VR journalism (Wu et al., 2021) and to empowering the audience.

The use of VR in journalism has highlighted some ethical conundrums, which apply to other forms of technology-driven innovation in the field. For instance, it raises issues of transparency: how much of VR journalism is manufactured while maintaining the illusion of 'showing a situation as is' (Aitamurto, 2018, p. 5)? This further relates to the notion of accuracy and objectivity: VR journalism has highlighted the infeasibility of the concept of objectivity, while again strengthening the importance of transparency (Sánchez Laws, 2019). Moreover, it sheds light on the discussion on the vulnerability and well-being of the audience.

The entrance of VR into the world of journalism was accompanied by hopes of re-engaging a disengaged audience by creating empathy, but also with hopes of improving the branding and economic viability of journalism agencies. While these hopes have only partially manifested, VR journalism still expands on the idea of what journalism is and does. In the context of VR journalism, stories are no longer simply told but experienced.

References

Ahn, S.J., Bostick, J., Ogle, E., Nowak, K.L., McGillicuddy Kara T. and Bailerson, J.N. (2016) 'Experiencing nature: Embodying animals in immersive virtual environments increases inclusion of nature in self and involvement with nature,' *Journal of Computer-Mediated Communication*, 21(6), pp. 399–419.

Aitamurto, T. (2015) 'Crowdsourcing as a knowledge search method in digital journalism: Ruptured ideals and blended responsibility,' *Digital Journalism*, 4(2), pp. 280–297.

Aitamurto, T. (2018) 'Normative paradoxes in 360-degree journalism: contested accuracy and objectivity,' *New Media and Society*, 21(1), pp. 3–19.

AP-VR360. (2016) 'Nimrud's riches: The islamic state's efforts to erase history' [Video]. Available at: https://www.youtube.com/watch?v=ZOTlGlWXB5g (Accessed: 28 January 2022).

Archer, D. and Finger, K. (2018) 'Walking in another's virtual shoes: Do 360-degree video news stories generate empathy in viewers?,' *Tow Centre for Digital Journalism*, Available at: https://doi.org/10.7916/D8669W5C (Accessed: 28 January 2022).

Aronson-Rath, R.; Milward, J.; Owen, T. and Pitt, F. (2016) *VR Journalism*. Tow Centre for Digital Journalism. Available at: https://towcentre.gitbooks.io/virtual-reality-journalism/content/ (Accessed: 28 January 2022).

Barreda-Ángeles, M., Aleix-Guillaume, S. and Pereda-Baños, A. (2020) 'VR storytelling as a double-edged sword: Immersive presentation of nonfiction 360-degree-video is associated with impaired cognitive information processing,' *Communication Monographs*, 88(2), pp. 1-20.

Buji , M. and Hamari, J. (2020a) 'Immersive journalism: Extant corpus and future agenda," *GamiFIN Conference 2020*, pp. 136–145.

Buji , M. and Hamari, J. (2020b) 'Satisfaction and willingness to consume immersive journalism: Experiment of differences between VR, 360 video, and article,' *ACM International Conference Proceeding Series*, pp. 120–125.

CNN (2016) 'Search & rescue after a deadly earthquake' [Video]. Available at: https://edition.cnn.com/2017/03/05/vr/italy-earthquake-2016-vr/index.html (Accessed: 28 January 2022).

Davis, D. P., & Millet, B. (2020) 'Designing 360 video for immersive journalism,' *Proceedings of the Human Factors and Ergonomics Society Annual Meeting*, 64(1), pp. 1059–1063.

de Bruin, K., de Haan, Y., Kruikemeier, S., Lecheler, S. and Goutier, N. (2020) 'A first-person promise? A content-analysis of immersive journalistic productions,' *Journalism* [online]. Available at: https://doi.org/10.1177/1464884920922006 (Accessed: 28 January 2022).

de la Peña, N., Weil, P., Llobera, J., Giannopoulos, E., Pomés, A., Spanlang, B., Friedman, D. (2010) 'Immersive journalism: Immersive VR for the first-person experience of news,' *Presence: Teleoperators and Virtual Environments*, 19(4), pp. 291–301.

Deuze, M. (2005) 'What is journalism? Professional identity and ideology of journalists reconsidered,' *Journalism*, 6(4), pp. 442–464.

Diemer, J., Alpers, G.W., Peperkorn, H.M., Shiban, Y. and Mühlberger, A. (2015) 'The impact of perception and presence on emotional reactions: A review of research in VR,' *Frontiers in Psychology*, 6(26), [online]. Available at: https://www.ncbi.nlm.nih.gov/pmc/articles/PMC4311610/ (Accessed: 28 January 2022).

Domínguez, E. (2017) 'Going beyond the classic news narrative convention: The background to and challenges of immersion in journalism,' *Frontiers in Digital Humanities*, 4(1), pp. 1–11.

Dowling, D. O. (2020) 'Place-based journalism, aesthetics, and branding', in: Uskali, T., Gynnild, S., Jones, S. and Sirkkunen, E. (ed.) *Immersive Journalism as Storytelling – Ethics, Production and Design*. New York: Routledge, pp. 99-112.

Hanitzsch, T. (2007) 'Deconstructing journalism culture: Toward a universal theory,' *Communication Theory*, 17(4), pp. 367–385.

Harris, J. and Taylor, J. (2021) 'Narrative in VR journalism: research into practice,' *Media Practice and Education*, 22(3), pp. 221-224.

Herrera, F., Bailenson, J., Weisz, E., Ogle, E. and Zaki, J. (2018) 'Building long-term empathy: A large-scale comparison of traditional and VR perspective-taking,' *Plos One*, 13(10), pp. 1–22.

Ilvonen, I., Vanhalakka, J., Helander, N. (2020) 'Case study: creating business value in immersive journalism,' in Uskali, T., Gynnild, S., Jones, S. and Sirkkunen, E. (ed.) *Immersive Journalism as Storytelling – Ethics, Production and Design*. New York: Routledge, pp. 112 - 122.

Jones, S. (2017) 'Disrupting the narrative: immersive journalism in VR,' *Journal of Media Practice*, 18(2), pp. 171–185.

Kang, S., O'Brien, E., Villarreal, A., Lee, W. and Mahood, C. (2019) 'Immersive journalism and telepresence: Does VR news use affect news credibility?,' *Digital Journalism*, 7(2), pp. 294–313.

Kool, H. (2016) 'The ethics of immersive journalism: A rhetorical analysis of news storytelling with VR technology,' *Intersect*, 9(3), pp. 1–11.

Mabrook, R. (2021) 'Between journalist authorship and user agency: Exploring the concept of objectivity in VR Journalism,' *Journalism Studies*, 22(2), pp. 209–224.

Maschio, T. (2017) 'Storyliving: An ethnographic study of how audiences experience VR and what that means for journalists,' *Google News Lab* [online]. Available at: https://newslab.withgoogle.com/assets/docs/storyliving-a-study-of-vr-in-journalism.pdf (Accessed: 28 January 2022).

Milk, C. (2015) 'Chris Milk: How VR can create the ultimate empathy machine' | TED Talk [online.]. Available at: https://www.ted.com/talks/chris_milk_how_virtual_reality_can_create_the_ultimate_empathy_machine/transcript (Accessed: 28 January 2022).

Newman, N., Fletcher, R., Schulz, A., Simge, A. and Nielsen, R.K. (2020) 'Journalism, media, technology trends and predictions 2020,' *Reuters Institute Digital News Report* [online]. Available at: http://www.digitalnewsreport.org/publications/2020/journalism-media-and-technology-trends-and-predictions-2020/ (Accessed: 28 January 2022).

Nielsen, S.L. and Sheets, P. (2019) 'Virtual hype meets reality: Users' perception of immersive journalism,' *Journalism*, 3, pp. 2637–2653.

Palmer, L. (2020) 'Breaking Free' from the frame: International human rights and the New York Times' 360-degree video journalism,' *Digital Journalism*, 8(3), pp. 386-403.

Rebenitsch, L., and Owen, C. (2016) 'Review on cybersickness in applications and visual displays,' *VR*, 20(2), pp. 101–125.

Rodríguez-Fidalgo, M.I. and Paíno-Ambrosio, A. (2020) 'Use of VR and 360-degree video as narrative resources in the documentary genre: Towards a new immersive social documentary?,' *Catalan Journal of Communication and Cultural Studies*, 12(2), pp. 239–253.

Ryfe, D. (2020) 'Journalism and Democracy,' in Wahl-Jorgensen, K. Hanitzsch, T. (ed.) *The Handbook of Journalism Studies*. New York: Routledge, pp. 293–306.

Sánchez Laws, A.L. (2019) *Conceptualising Immersive Journalism*. New York: Routledge.

Schutte, N.S. and Stilinović, E.J. (2017) 'Facilitating empathy through VR,' *Motivation and Emotion*, 41(6), pp. 708–712.

Shin, D. (2018) 'Empathy and embodied experience in virtual environment: To what extent can VR stimulate empathy and embodied experience?' *Computers in Human Behavior*, 78, pp. 64–73.

Shin, D. and Biocca, F. (2018) 'Exploring immersive experience in journalism,' *New Media and Society*, 20(8), pp. 2800–2823.

Sirkkunen, E., Vazquez-Herrero, J., Uskali, T. and Väätäjä, H. (2020). 'Exploring the immersive journalism landscape,' in Uskali, T., Gynnild, S., Jones, S. and Sirkkunen, E. (ed.) *Immersive Journalism as Storytelling – Ethics, Production and Design.* New York: Routledge, pp. 13–24.

Shoemaker, P., and S. Reese. (2014). *Mediating the Media Message in the 21st Century: A Media Sociology Perspective.* 3rd edition. New York: Routledge.

Slater, M., Navarro, X., Valenzuela, J., Oliva, R., Beacco, A., Thorn, J. and Watson, Z. (2018) 'Virtually being Lenin enhances presence and engagement in a scene from the Russian Revolution," *Frontiers in Robotics and AI*, 5(91), [online]. Available at: https://doi.org/10.3389/frobt.2018.00091 (Accessed: 28 January 2022).

Society of Professional Journalists (2014) *SPJ Code of Ethics.* [online]. Available at: http://www.spj.org/ethicscode.asp (Accessed: 28 January 2022).

Steed, A., Pan, Y., Watson, Z. and Slater, M. (2018) "We wait'-The impact of character responsiveness and self embodiment on presence and interest in an immersive news experience,' *Frontiers in Robotics and AI*, 5(10), pp. 1–14.

Steinfeld, N. (2020) 'To Be there when it happened: Immersive journalism, empathy, and opinion on sexual harassment," *Journalism Practice*, 14(2), pp. 240–258.

Sundar, S. S., Kang, J. and Oprean, D. (2017), "Being There in the Midst of the Story: How Immersive Journalism Affects Our Perceptions and Cognitions." *Cyberpsychology, Behavior and Social Networking*, 20(11), pp. 672-682.

Tuchman, G. (1972) 'Objectivity as strategic ritual: An examination of newsmen's notions of objectivity,' *American Journal of Sociology*, 77(4), pp. 660–679.

Van Damme, K., All, A., De Marez, L. and Van Leuven, S. (2019) '360-degree video journalism: Experimental study on the effect of immersion on news experience and distant suffering,' *Journalism Studies*, 20(14), pp. 2053–2076.

Vettehen, H. P., Wiltink, D., Huiskamp, M., Schaap, G. and Ketelaar, P. (2019) 'Taking the full view: How viewers respond to 360-degree video news,' *Computers in Human Behavior*, 91, pp. 24–32.

Vindenes, J. and Gynnild, A. (2020) 'Case Euronews: a low-cost approach to immersive storytelling,' in Uskali, T., Gynnild, S., Jones, S. and Sirkkunen, E. (ed.) *Immersive Journalism as Storytelling – Ethics, Production and* Design. New York: Routledge. pp. 25-36.

Wang, G., Gu, W. and Suh, A. (2020) 'The effects of 360-degree VR videos on audience engagement: Evidence from the New York Times,' in Nah, F. and Xiao, B. (ed.) *International Conference on HCI in Business, Government, and Organisations.* Las Vegas:Springer, pp. 217–235.

Watson, Z. (2017) 'VR for news: the new reality?,' *Digital News Publications,* [online]. Available at: http://www.digitalnewsreport.org/publications/2017/vr-news-new-reality/ (Accessed: 28 January 2022).

Wu, H., Cai, T., Liu, Y., Luo, D. and Zhang, Z. (2021) 'Design and development of an immersive VR news application: a case study of the SARS event,' *Multimedia Tools and Applications*, 80, pp. 2773–2796.

Zelizer, B. (2012) 'About to die: How news images move the public,' *Mass Communication and Society*, 15(5), pp. 782–785.

Chapter 5

Table for Two - A parallel interactive narrative in VR (PIN VR)

Delwyn Jude Remedios

Indian Institute of Technology Hyderabad, India

Deepak John Mathew

Indian Institute of Technology Hyderabad, India

Max Schleser

Swinburne University

Abstract

Parallel interactive narratives in VR (PIN VR) offers new ways for audiences to interact with narratives in cinematic VR. Cinema is constantly evolving through narrative innovations and emerging technologies continue to offer new explorations in storytelling. This chapter applies a practice-based research method to investigate the production process for a PIN VR artefact *Table for Two. Table for Two* experiments with multiple genres in a one-shot PIN VR experience, captured simultaneously with three VR cameras. *Table for Two - A Parallel Interactive Narrative in VR (PIN VR)* outlines new insights for narrative design in cinematic VR.

Keywords: Parallel Interactive Narratives in VR, PIN VR, Cinematic VR, multi-genre, practice-based research

***

Introduction

Table for Two is a short parallel interactive narrative in VR (PIN VR) that explores multiple genres in an urban café. Selected genres such as romance, drama and supernatural fiction are seamlessly integrated in the same environment. This experience is designed to encourage the navigator to teleport between genres. Research on *Cinema 3.0* suggests that the role of a passive film viewer has transformed into a navigator with the increase of interaction (Daly, 2008, pp. 157-165). This chapter asserts that the term 'navigator' is appropriate for an experience such as *Table for Two*, due to their active role within the VR environment. As a navigator teleports in *Table for Two*, the time lapses in each parallel narrative. In conventional cinema, parallel narratives are commonly shown in different timelines of a film, whereas PIN VR provides a platform for each parallel narrative to occur simultaneously on the cinematic VR timeline. While a PIN VR experience will conclude at the same time for each navigator, they will undergo a unique sequential order, based on their teleportation choices. *Table for Two* further encourages a navigator to engage with multiple viewings of the experience to unfold the interconnections between the different narratives.

In this chapter, PIN VR is proposed as an alternative approach for narratives in VR. Parallel narratives in cinema provide a unique way of depicting the chaos and randomness of human life (Roth, 2010). Abbot (2008) states that these narratives replicate the world we live in. Parallel narratives has been addressed by a variety of names such as 'network narrative' (Bordwell, 2006b), 'hyperlink cinema' (Quart, 2005) and 'smart film' (Peters, 2008). Aronson has extensively researched parallel narrative in her book *21st Century Screenplay: A comprehensive guide to writing tomorrow's film*. Here, she describes parallel narrative as "several separate narratives running in parallel, often involving non-linearity, time jumps, large casts, or all of these" (Aronson, 2010, p. 167).

These complex narrative structures have challenged the cinema viewers to unfold non-linearity and identify networks within parallel plots. While parallel narratives have been used in stage theatre, early cinema, and television serials (Roth, 2010), they are becoming increasingly popular in mainstream cinema (Bordwell, 2006b). It is observed that parallel narrative in cinema has similarities with *montage* theory (Bordwell, 1972) in terms of fragmentation. Booker (2007) mentions that the increase of fragmentation in postmodern films can be seen as an extension of older montage techniques and therefore an evolution of film as a medium. This leads to the inquiry if mainstream cinema is undergoing yet another reinvention through the progression of parallel narrative (Roth, 2010)?

Though cinema is constantly evolving through narrative innovations, emerging technologies continue to offer new explorations. It is evident that VR is one of the latest technologies in cinema (Mateer, 2017) and parallel narratives are gradually emerging in the context of this technology. An advantage of VR is that the viewer can step into the borders of the cinema screen and explore a 360-degree environment. This radical transformation has challenged filmmakers to explore new approaches to VR narratives. While some research on VR delves into the grammar employed to guide viewers towards intended areas in a VR environment (Pillai and Charles, 2017; Sheikh, Brown and Watson, 2016), Gabler, Siegmund and Braun (2018) argue that VR should empower the viewer with exploratory freedom to utilize its full potential. In response, this research argues that the freedom of the navigator to choose her viewing direction can be addressed by PIN VR. This chapter observes that parallel narratives in VR lack experimentation in terms of the design and production processes. This leads to questioning what is the design and production process for a PIN VR experience? A practice-based research method is applied to discuss the production process applied for *Table for Two* as a case study of PIN VR. Immersive and interactive experiences such as cinematic VR (CVR), seamless multithread films and *cinévoqué* are studied to position PIN VR on a scale of interactivity. The concept of PIN VR and the interactive experience of the navigator is further explained in the context of *Table for Two*. This is followed by the theoretical framing of parallel narratives, along with its variations and challenges in cinema and VR. The chapter will further discuss the type of parallel narrative structure observed in *Table for Two*.

Table for Two - Design and production process

Table for Two experiments with multiple genres such as romance, drama and supernatural fiction in a one-shot PIN VR experience. This experience is captured simultaneously with three VR cameras located within an urban café. In the writers' room, it was discussed that the space of a café provides an opportunity to represent multiple genres within conversations across tables. This experience involves seven main characters; three pairs of guests and a waiter who interacts with all characters in the café. A summary of the characters and narratives in each genre is further described.

In urban India, it is commonly observed that couples meet at a café. *Table for Two* depicts an encounter of a young couple meeting after the COVID-19 lockdown. For example, Sakshi and Rohan had initially met at an inter-college sports event. They discovered their interest in each other through online chatting; however, they had not met in person after the sports event. Sakshi is a final year student residing in Delhi and Rohan has recently started working in Mumbai. Sakshi had come to Mumbai for her cousin's engagement party and was interested in meeting Rohan during the few days that she was in the city. Their conversation at the café plays a significant role in unfolding their relationship.

In other cases, users experienced a supernatural fiction narrative in *Table for Two*. The term supernatural refers to elements that cannot be explained by rational or scientific means (Rankin, 2011). In one example, the character Neha had recently encountered tragic visions that took place in reality at a later time.

She was very uncomfortable with these repeated visions and had confided in her close friend Simran, who has experienced such activities. As Neha narrates the events to another character, she is interrupted with yet another vision predicting a tragic event in the café. Ramirez states that supernatural fiction uses the element of suspense to build tension (Ramirez, 2018). Therefore, the plot for this particular narrative appropriately uses suspense.

Drama relies on the emotional and relational development of realistic characters (Buffam, 2011). *Table for Two* employs drama as a genre to balance out the contrast between the romance and supernatural fiction. In one iteration of *Table for Two*, a young photographer Harjyot, had recently lost his job during the COVID-19 pandemic. He was looking for freelance opportunities to showcase his skill and earn his living. Harjyot had been selected to design the cover page for an interior design magazine. He has a meeting with his client Vikram in the café. In one scenario, Harjyot presents his concepts to another character who is not interested in Harjyot's work. Drama is driven by the expectation for Harjyot to succeed in his presentation. The application of multiple genres in the same experience led to challenges in the script, storyboard and set design process.

Dooley reports various scriptwriting approaches of three Australian VR practitioners, Mike Jones, Tom Phillips and Lynette Wallworth. Her study points out that there is no standard approach for designing a VR script (Dooley, 2018). Similarly, *Table for Two* followed customised scriptwriting explorations resulting in a rotating cylindrical PIN VR script display. Initial explorations included a traditional format, a tabular format and a spherical display. The *Table for Two* crew members' inputs led to a rotating cylindrical PIN VR script display. Figure 5.5 shows the crew interacting with the display during a rehearsal session. They mentioned that display helped them to understand their roles and connect the timelines of each parallel narrative effectively. As per the recommendations of the crew, this display was utilized during rehearsals and the shoot.

In his article *Storyboarding in 360°*, Merkle (2017) highlights that storyboards for cinematic VR are not limited to creating frames to guide the viewer, instead create a world to discover the story. Therefore, a cinematic VR storyboard should consider everything that happens in the story world. Like the online discussion platforms, the storyboard for *Table for Two* was approached in a split-screen format as shown in figure 5.1. There were three separate versions, each highlighting the occurrences at any one table in a larger size as compared to the other tables. This method was effective in helping the crew understand their roles, expected gestures, and overall plot. Along with the script and storyboard, the set design also plays a key role in a cinematic VR narrative.

Figure 5.1: Table for Two storyboard presented in split-screen format.

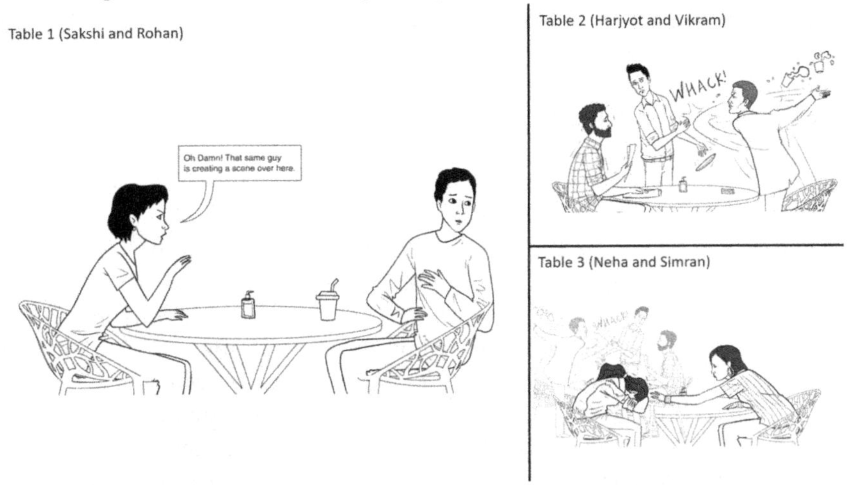

Jenkins (2003) suggests that interactive experience designers can adopt the techniques of environmental storytelling, found in designing amusement parks. In amusement parks design, the narrative element is infused into the physical space as a guest navigates through the set. This recommendation is useful for planning a cinematic VR narrative. In *Table for Two*, the space of the café plays a critical role in preparing the navigator to interact with the narratives. As the experience represents multiple genres, the set design for each genre had to convey the mood seamlessly in the café. Lighting and props were customized as per the genres. Indian commercial cinema, also known as Bollywood, was selected as the overall theme of the café. Popular film posters were framed and displayed keeping in mind that the genres of the posters corresponded to the genre of the narrative. In conventional cinema, the production equipment can be accommodated outside the borders of the shooting area; however, in cinematic VR, the production equipment is required to be a part of the set as everything is visible to the navigator. Customised spotlights and table lamps were used to light the cast, while white table maps were carefully selected as light reflectors.

Table for Two describes the conceptual design challenges and new production methods such as the rotating cylindrical PIN VR script display, the split screen storyboard format and incorporation of production equipment in the set design. The concept of PIN VR is further explained in the following section.

Parallel interactive narratives in VR (PIN VR)

Table for Two is a short narrative experience that is a case study of PIN VR. A navigator in a PIN VR can seamlessly teleport from one narrative location to another, while the remaining narrative locations run in parallel sequences, irrespective of the navigator's presence. This interaction takes place by gazing at 'pinpoint' areas in a VR environment. The unique feature of PIN VR is that the navigator's progression results in a time lapse in each parallel scene. In *Table for Two*, the navigator is given the freedom to choose the genre of interest and can seamlessly switch from one genre to the next within the timeline of the experience. As the narratives are interconnected, the navigator is left with missing links, and is encouraged to re-navigate the experience to fix the pieces of the puzzle together. The navigator additionally plays the role of the editor as their gaze will determine the narrative sequence. The role of the film director in PIN VR is to design the destiny for the narrative world. The director's decisions will determine the occurrences in each narrative, while the editing freedom is handed over to the navigator. Parallel narratives in conventional cinema are presented with the aid of cuts and editing techniques within the timeline of a film (Paul, 2016). Hodgkinson (2019) argues that one of the most disorienting experiences in VR is the camera cut. In VR, the navigator may have no fixed visual references, therefore, any sudden change can cause disorientation. However, the limitations of cuts and editing techniques in VR can be used as an advantage to represent parallel narratives simultaneously in a VR environment. The immersive and interactive component of PIN VR is further described in comparison to cinematic VR, seamless multithread films and *cinévoqué* experiences.

Mateer (2017) describes cinematic VR (CVR) as a type of immersive VR experience where individual navigators can look around a 360-degree synthetic world with stereoscopic views and spatial audio. A CVR experience limits the navigator to choose viewpoints within the experience, rather than interacting with the virtual world itself. Mateer argues that there is a lack of research on formal strategies to be used in developing a CVR experience.

Garcia et al. (2017) propose a seamless multithread experience system for the production of VR narratives that permit a navigator to seamlessly interact with the content. The navigator makes subliminal decisions by focusing their attention at predefined areas. This function provides a seamless opportunity to choose the path of the story that they are currently engaged with. In a seamless multithread experience, the navigator is not aware that they have made a choice and the immersion remains uninterrupted. An alternate experience corresponding to seamless multithread film experience is found in a *cinévoqué* experience.

Pillai, Murugan, and Dev (2019) define *cinévoqué* as a narrative experience that dynamically responds to passive interactions from the navigator, unknown to them. The term *cinévoqué* is derived from cinema *évoqué* (evoked cinema). This experience differs from interactive cinema that allows the navigator to make conscious choices about the progression of the narrative. Here, it dynamically modifies the narrative and other elements within the environment based on the navigator's gaze. On a scale of interactivity, a *cinévoqué* is positioned between an immersive CVR film and an interactive, immersive VR experience. The concept of *cinévoqué* has been further explained through two short films, *Schrödinger's Vada-Pav* and *Shapeshifter. Shapeshifter* explored the possibility of varying the genre of the experience across storylines. While the experience branches into a comedy or a thriller, both these genres are connected to science fiction. They recommend that the possibilities of each narrative sequence leading to a unique genre could be further explored. Therefore, *Table for Two* explores multiple genres.

The readings of CVR, seamless multithread films and *cinévoqué* reflect the following observations. CVR is immersive and not interactive, while seamless multithread films and *cinévoqué* are immersive plus interactive. In a seamless multithread film or a *cinévoqué,* the navigator makes choices unaware to them, that alter the outcome of the narrative. In a PIN VR experience, the navigator seamlessly makes interactive choices; however, these choices do not modify the outcome of the narrative. Therefore, this study positions PIN VR on a scale of interactivity in figure 5.3.

Figure 5.2: PIN VR experience on a scale of interactivity.

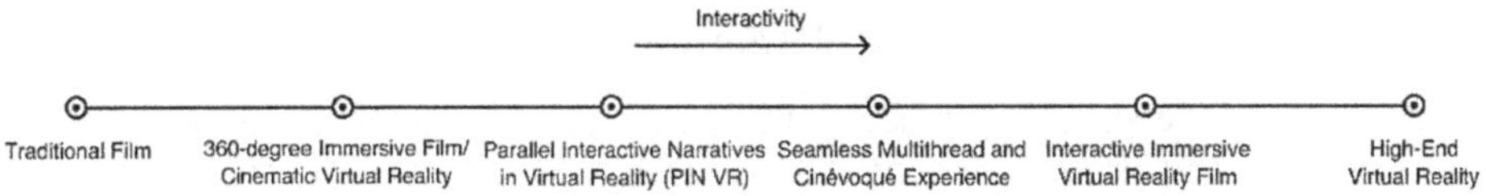

Let's assume the navigator starts the experience in narrative one. After one minute, she teleports to narrative two with a seamless gaze. At five minutes, she teleports to narrative three and at seven minutes she returns to narrative one. The teleportation time period can change based on the navigator's choices. From this figure, it is understood that all narrative timelines will lapse, irrespective of where the navigator is currently immersed. The navigator's choices in PIN VR unfold alternative narratives and can lead to an infinite narrative possibility. The following section provides a theoretical framework for understanding more about parallel narratives and its variations in cinema.

Parallel narrative in cinema

Bordwell (2006a) defines parallel narrative as a network narrative which highlights several protagonist inhabiting distinct, but intermingling storylines. Building on this understanding, Denton (2020) states that parallel narratives in cinema include two or more separate narratives linked by a common character, event, or theme. In *Table for Two,* each genre represents a separate narrative, which is placed in the common environment of a café, amidst the event of the COVID-19 pandemic. Denton further mentions that parallel narratives generate new narratives through symbolic parallels and associations, due to which infinite parallels can emerge from this creative process. The previous section on PIN VR explained that its network structure can lead to infinite narratives. Parallel narratives are further explained with examples in the context of cinema.

Parallel narratives have been observed in early cinema, such as *The Kleptomaniac* (Porter, 1905). This film delivers a social message about the contrasting fates of the rich and poor. Another significant innovation in cinematic parallel narrative can be found in the film *Rashomon* (Kurosawa, 1950), which experimented with non-linear narratives by describing a common event with multiple points of view. A further breakthrough in cinematic parallel narrative can be seen in the film *Pulp Fiction* (Tarantino, 1994). In this film, three plots overlap in terms of characters, events and time. Cameron (2008) describes *Pulp Fiction* as an *anachronic* narrative structure. *Anachronic* narratives refer to a modular structure that modifies a flashback by removing the hierarchy between the primary and anterior narrative temporality. These examples of parallel narratives in cinema demonstrate that parallel narratives have variations in form and application. This study further investigates Aronson's classification of cinematic parallel narrative.

Types of parallel narratives in cinema

Aronson (2010c) classifies parallel narratives into two main categories, ensemble films and time jumps. Ensemble films usually involve large casts, multiple storylines and linearity, without time jumps; whereas time jumps films involve non-linearity and fractured storylines. Ensemble films can be further classified as tandem narratives, multiple protagonist narrative and double journey narrative. Time jump films can be further classified as flashback, consecutive stories and fractured tandem. While each of these forms offers a unique representation of parallel narrative, Aronson states that there is a possibility to encounter interconnections within these forms. The different types of ensemble films and time jump films are further highlighted.

Ensemble films

Tandem narratives are ensemble films that refer to equally weighted stories running simultaneously. The term 'in tandem' suggests events taking place at the same time. Most characters in a tandem narrative are affected by the same socio-political event. These films, which include examples like *City of Hope* (Sayles, 1991) and *Traffic* (Soderbergh, 2000), are usually truncated and are held together by a macro plot. Multiple protagonist narratives represent groups rather than an individual's journey. These are films about a small group in a quest, a reunion or an emotional siege. Often films about families are considered multiple protagonist films as they portray emotional sieges. E.g. *Little Miss Sunshine* (Dayton and Faris, 2006) and *The Full Monty* (Cattaneo, 1997). Double journeys depict two equally important protagonists who are journeying either towards, apart or in parallel with each other physically, emotionally, or both. Examples include *Finding Nemo* (Stanton, 2003) and *The Departed* (Martin Scorsese, 2006).

Time jump films

Flashbacks are a form of time jump films that can take place through backstories, regrets, bookends, previews, life-changing incidents and double narrative flashbacks. Examples of such films include *Fight Club* (Fincher, 1999) and *Slumdog Millionaire* (Boyle, 2008). Consecutive stories refer to equally weighted stories following one after the other and joined together at the end (e.g., *Run Lola Run* (Tykwer, 1998) and *The Butterfly Effect* (Bress and Gruber, 2004)). Fractured tandem consists of equally-weighted stories, often in different time frames, fractured, truncated and put together to add suspense, and a few notable examples are *Babel* (Iñárritu, 2006) and *The Hours* (Stephen Daldry, 2002).

In her chapter *Introduction to Parallel Narratives*, Aronson (2010b) states that the rearrangement of a linear narrative into a non-linear model can create excitement and suspense. However, Aronson also argues that parallel narrative films also struggle with pace, connection, meaning and closure because they have so many plotlines. Therefore, a writer who rejects Aristotle's three-act structure is instantly challenged with the problem of finding new methods to tell a parallel narrative. Homer's *Odyssey*, which was written down centuries before Aristotle began at a crucial scene in the middle of the story and then used flashbacks. Aronson states that Homer's *Odyssey* is closer to *Pulp Fiction* and *21 Grams* than it is to the three-act paradigm as it starts in the middle, uses multiple protagonists, and tells Odysseus' adventures after leaving Troy in a flashback.

In *Table for Two*, the experience begins with the couple representing the romance genre, in the middle of a conversation. The photographer is simultaneously observed waiting for the client to arrive. The drama narrative takes action as the client steps into the café. After which, the supernatural genre characters enter the café. This implies that there are three parallel conversations currently present in the café. This structure reveals that each narrative has its own starting and ending point. Aronson mentions that there is a lack of film theory available for writers who want to use parallel narratives structures. Similarly, VR lacks theoretical knowledge in planning and developing VR experiences (Louchart and Aylett, 2003). As a result, the production process of *Table for Two* addresses this concern.

Emergence of parallel narratives in VR

Various observations of parallel narratives in VR discourse the navigator's interaction within these narratives. Martin and Mills (2019) describe a 360-degree tandem parallel narrative film that depicts multiple conversations. This experience aims to test the navigator's comprehension, engagement, and response as they engage with the characters. However, the demonstration and explanation of the concept is currently in a development phase. Google's 'Advanced Technology & Projects (ATAP)' animated short film *Duet* (Keane, 2014) is an example of a double journey parallel narrative in VR. This experience gives the navigator multiple possibilities to seamlessly choose between two storylines. If the navigator is lost in the VR environment, the experience will come to a halt; while a character such as a butterfly will lead them back to one of the characters and only then, the narrative will continue. A VR experience *New Wave* (Mallal and Hjartarson, 2015) is another example of a double journey parallel narrative, which depicts a couple at a beach, mentally analysing their relationship. The navigator's gaze enables them to hear the inner thoughts of a specific character through spatial audio. While the 360-degree video remains the same, each navigator witnesses a different experience based on the customized audio. For that reason, the navigator can ostensibly experience a different narrative (Begel, 2016). *Mr. Robot VR* (Esmail, 2016) has explored flashback narratives by engaging the navigator on a journey with the lead character, as he recalls an encounter with his love interest (Bishop, 2016). A Liberty Mutual Insurance Advertisement portrays a 360-degree video quiz where the spherical space is divided into four sections by splitting the screen; a title and three concurrent mini-stories that depict alternative courses of action. The navigator has the freedom to select a scenario of their choice, which will unfold that particular outcome (Levine, 2015). While this is a split-screen representation of parallel narratives; it can convert into a consecutive story, if the navigator decides to explore one option and then try the others.

These examples explain how Aronson's forms of cinematic parallel narratives can be adopted into VR. These experiences reveal that there are limited experiments using parallel narratives in VR and this area could widen with an increase in experimentation. While these examples successfully explore parallel narratives, they have not explored the possibility of navigator teleportation. We argue that the navigators' teleportation can further increase the experimentation of parallel narrative structures. The examples also highlight that there is a lack of understanding about the design and production processes of parallel narratives in VR. These findings have led the researcher to experiment with PIN VR as an alternative approach for narratives in VR, while emphasising the production process of *Table for Two*.

In the context of PIN VR, this study recommends that tandem and multiple protagonist narratives are most appropriate for exploring PIN VR, as these forms can branch out into multiple narratives. *Table for Two* is an example of a hybrid between a tandem and multiple protagonist narrative. It is considered a tandem narrative as each genre in the café is equally important. These conversations are taking place at the same time and at the same location, while each character is affected in some way by the COVID-19 pandemic. Each of the main characters in the experience can be considered a protagonist. The navigator' choices will decide which character's journey to follow. Therefore, the experience is also considered a multiple protagonist narrative. A double journey narrative additionally has potential in PIN VR, however, it is limited to two narratives. As PIN VR experiences are running simultaneously in real time, a fractured tandem and flashbacks have limitations in time jumps. Time jumps in consecutive stories suggest when one story ends, the next one begins. While a PIN VR timeline limits this form, it can be achieved if the navigator experiences one narrative without teleportation, re-experiences the next narrative and so on. Thus, there is scope to further experiment with the variations of parallel narratives in PIN VR.

Conclusion

A discussion on VR experiences explains how Aronson's types of parallel narratives can be presented in VR. These experiences have unique functional qualities; however, they lack exploration in teleportation and insights about the production process. This chapter addresses these limitations with the concept of parallel interactive narratives in VR (PIN VR) with a practise-based artefact *Table for Two.*

PIN VR offers a navigator the freedom to teleport from one narrative location to the other while the remaining narratives continue in parallel, irrespective of their presence. This experience differs from cinematic parallel narratives as the narratives are often shown at different points in the cinema timeline, whereas PIN VR presents the parallel narratives simultaneously. The role of the director of a PIN VR is to design the narrative world, while the role of the navigator is to determine the editing sequence based on their teleportation choices. Therefore, allowing the navigator to customize their experience as they desire.

Table for Two is a hybrid between a tandem and multiple protagonist narrative as it interconnects three different narrative genres of equal importance. A café was selected as a suitable location to represent multiple genres seamlessly in the same environment, thus highlighting the importance of location in PIN VR. However, this led to challenges in the script, storyboard and set design. The challenge of the script was resolved through a rotating cylindrical PIN VR script display. This display allowed the crew members to connect the timelines of the parallel scripts, thus understanding their role effectively. Similarly, a spilt screen format was adopted for the storyboard, and the set was designed to give a cue to the genres. The production equipment was further integrated as props within the café. Therefore, the new production processes involved in designing *Table for Two* has resulted in new ways for audiences to build customised narrative experiences through PIN VR.

References

Abbot, H. P. (2008) *The Cambridge introduction to narrative.* 2nd edn. New York: Cambridge University Press.

Aronson, L. (2010a) *21st century screenplay: A comprehensive guide to writing tomorrow's film.* New York: Allen & Unwin, p. 167.

Aronson, L. (2010b) 'An introduction to parallel narrative,' in *21st century screenplay : A comprehensive guide to writing tomorrow's films.* New York: Allen & Unwin, pp. 167–170.

Aronson, L. (2010c) 'Six categories of parallel narrative,' in *21st century screenplay : A comprehensive guide to writing tomorrow's films.* New York: Allen & Unwin, pp. 171–176.

Babel (2006) Directed by Alejandro González Iñárritu [Film] United States, Mexico, Morocco, France: Paramount Vantage (North and Latin America) Mars Distribution (France), Summit Entertainment (International).

Begel, P. (2016) *Storytelling in a 360° Space* [online]. Available at: https://philipbegel.super.site/all-projects/storytelling-in-a-360-space#01272ae949f54bddb0882eaddd6b492b (Accessed: 18 August 2019)

Bishop, B. (2016) *The Mr. Robot VR Experience isn't a gimmick - It's true storytelling* [online]. Available at: https://www.theverge.com/2016/7/22/12253358/mr-robot-virtual-reality-elliot-flashback-story-sdcc-2016 (Accessed: 27 November 2021)

Booker, M. K. (2007) *Postmodern Hollywood: What's new in film and why it makes us feel so strange.* New York: Praeger Publishers.

Bordwell, D. (1972) 'The Idea of Montage in Soviet Art and Film,' *Cinema Journal - The Journal of the Society for Cinema & Media Studies,* 11(2), pp. 9–17.

Bordwell, D. (2006a) *Observations on Film Art* [online]. Available at: https://www.davidbordwell.net/blog/2006/11/27/lessons-from-babel/ (Accessed. 25 January 2022)

Bordwell, D. (2006b) *The way Hollywood tells it: Story and style in modern movies.* 1st edn. Los Angeles: University of California Press.

Buffam, N. (2011) *Drama* [online]. Available at: https://thescriptlab.com/screenplay/genre/952-drama/ (Accessed: 11 December 2021)

Cameron, A. (2008) 'Modular narratives in contemporary cinema,' in *Modular Narratives in Contemporary Cinema,* London: Palgrave Macmillan. pp. 1–19.

Daly, K. M. (2008) *Cinema 3.0: How Digital and Computer Technologies are Changing Cinema.* Ph.D. Thesis. Columbia University. pp.157-165.

Denton, T. (2020) *Parallel Narratives* [online]. Available at: https://parallelnarratives.org/about/ (Accessed: 13 February 2021)

Dooley, K. (2018) 'Scripting the virtual: Formats and development paths for recent Australian narrative 360-degree VR projects,' *Journal of Screenwriting,* 9(2) pp. 1–26.

Duet (2014) Directed by Glen Keane [Cinematic Virtual Reality]. United States: Google's Advanced Technology & Projects.

Fight Club (1999). Directed by David Fincher [Film]. United States: 20th Century Fox.

Finding Nemo (2003) Directed by Andrew Stanton [Film]. United States: Buena Vista Pictures Distribution.

Gabler, F., Siegmund, D. and Braun, A. (2018) 'Cinematic Narration in VR – Rethinking Film Conventions for 360 Degrees Cinematic Narration in VR,' in: Chen, J. and Fragomeni, G. (eds) Virtual, Augmented and Mixed Reality: Applications in Health, Cultural Heritage, and Industry. *International Conference on Virtual, Augmented and Mixed Reality.* Lecture Notes in Computer Science, 10910. Springer, Cham.

Garcia, O. D. R., Tag, B., Ohta, N., Sugiura, K. (2017) 'Seamless multithread films in VR,' *TEI 2017 - 11th International Conference on Tangible, Embedded, and Embodied Interaction,* Yokohoma. pp. 641-646.

Hodgkinson, G. (2019) 'Cut, don't cut - moving the viewer through a story in animated VR,' *Confia - 7th International Conference on Illustration and Animation,* Viana do Castelo. pp. 50-57

Jenkins, H. (2003) *Game Design as Narrative Architecture* [online]. Available at: https://paas.org.pl/wp-content/uploads/2012/12/09.-Henry-Jenkins-Game-Design-As-Narrative-Architecture.pdf (Accessed: 04 January 2022)

Levine, B. (2015) *Liberty Mutual gets millions of views for parallel stories in a 360-degree video ad on Facebook* [online]. Available at: https://marketingland.com/liberty-mutual-gets-million-views-parallel-stories-360-degree-video-ad-facebook-173394 (Accessed: 15 April 2019)

Little Miss Sunshine (2006). Directed by Jonathan Dayton and Valerie Faris [Film]. United States: Fox Searchlight Pictures.

Louchart, S. and Aylett, R. (2003) 'Towards a narrative theory of VR,' *Virtual Reality,* 7(1), pp. 2–9.

Martin, K. and Mills, T. (2019) *Parallel Narratives in 360 Video* [online]. Available at: https://ivrpa.org/news/parallel-narratives-in-360-video/ (Accessed: 12 September 2019)

Mateer, J. (2017) 'Directing for Cinematic VR: how the traditional film director's craft applies to immersive environments and notions of presence,' *Journal of Media Practice,* 18(1), pp. 14–25.

Merkle, C. (2017) *Storyboarding in 360°* [online], Available at: https://medium.com/cinematicvr/storyboarding-in-360-2ddce59d627d (Accessed: 14 January 2022)

Mr. Robot VR (2016). Directed by Sam Esmail [Cinematic Virtual Reality]. United States: Jaunt VR, USA Network.

New Wave (2015). Directed by Samir Mallal and Aron Hjartarson. United States: Spoke, VRSE.works.

Paul, J. (2016) *Master the Hollywood Technique of Parallel Editing* [online]. Available at: https://www.premiumbeat.com/blog/parallel-editing-hollywood-way/ (Accessed: 14 October 2017)

Peters, L. (2008) 'Private Fears in Public Places: Network Narrative and the Post-'Smart' American Melodrama' in D'Aoust, A. (ed.) *Synoptique* [online]. 12th edn. pp. 5-11. Available at: https://www.synoptique.ca/_files/ugd/811df8_6b0227b7b3594d2faae6c80c37b4183a.pdf (Accessed: 4th June 2022)

Pillai, J. S., Ismail, A. and Charles, H. P. (2017) 'Grammar of VR Storytelling: Visual Cues,' *Virtual Reality International Conference - Laval Virtual 2017,* France. pp. 1-4.

Pillai, J. S., Murugan, A. and Dev, A. (2019) 'Cinévoqué: Design of a Passively Responsive Framework for Seamless Evolution of Experiences in Immersive Live-Action Movies,' in Lamas, D. et al. (ed.) *Human-Computer Interaction – INTERACT 2019.* Lecture Notes in Computer Science, 11746. Springer, Cham. pp 359–367

Pulp Fiction (1994). Directed by Quentin Tarantino [Film]. United States: Miramax Films.

Quart, A. (2005) 'Networked: Dysfunctional families, reproductive acts, and multitasking minds make for Happy Endings,' *Film Comment,* 41(4), pp. 48-51.

Ramirez, R. (2018) *Secrets of the Horror Genre* [online]. Available at: https://storygrid.com/secrets-of-the-horror-genre/ (Accessed: 12 December 2021)

Rankin, A. (2011) *What Is Supernatural Fiction?* [online]. Available at: https://www.masterclass.com/articles/guide-to-supernatural-vs-paranormal-fiction#what-is-supernatural-fiction (Accessed: 11 December 2021)

Rashomon (1950). Directed by Akira Kurosawa [Film]. Japan: Daiei Film.

Roth, J. (2010) *Life is Messy: An Exploration of Parallel Narrative.* Bachelor's thesis. RMIT University.

Run Lola Run (1998). Directed by Tom Tykwer. Germany: Prokino Filmverleih.

Sheikh, A., Brown, A., Watson, Z. and Evans, M. (2016) 'Directing attention in 360-degree video,' International Broadcasting Convention *2016 Conference.* Amsterdam.

Slumdog Millionaire (2008) Directed by Danny Boyle [Film]. United Kingdom: Pathé Distribution.

The Butterfly Effect (2004) Directed by Eric Bress and J. Mackye Gruber [Film]. United States: New Line Cinema.

The Departed (2006) Directed by Martin Scorsese [Film]. United States: Warner Bros. Pictures.

The Full Monty (1997) Directed by Peter Cattaneo [Film]. United Kingdom: Fox Searchlight Pictures.

The Hours (2002) Directed by Stephen Daldry [Film]. United States: Paramount Pictures, Miramax Films.

The Kleptomaniac (1905) Directed by Edwin S. Porter [Film]. United States: Edison Manufacturing Company.

Section 2: VR and nature

Chapter 6

VR's growing role in establishing environmental empathy and its potential to model complex climate systems

Mary Le Rouge

Cleveland Institute of Music

Abstract

VR has been used extensively to promote public empathy for the environment, especially for coral reefs in the face of bleaching caused by global warming. While empathy is important, VR has much greater potential to model climate systems and allow students and citizen scientists alike to help solve environmental problems. VR programs that model climate simulations and interventions using Gaia theory can offer a greater understanding of the Earth's functions and diseases, and a safe space to develop solutions that can forecast how human interventions in climate will play out in multiple systems over long time periods.

Keywords: VR; climate system models; medical simulations; Gaia theory; holistic medicine

***

Introduction

VR technology has seen a surge during the past 10 years. The technology has become more popular not only for entertainment, but also for education in a wide variety of contexts, including workplaces, schools, and museums. When used to educate about the dangers of climate change, VR has been shown to induce empathy for environmental issues and results in significant behavioural changes in the physical world toward environmental protection (Bailenson, 2016; Fauville et al., 2021). VR makes it possible for those who do not have the resources or physical ability to obtain physical-world experiences to still achieve a secondary experience in the classroom or other educational context.

One of the most popular types of VR experiences that tackles the problem of climate change is films that show coral reefs in all their abundance, and conversely, the damage to corals caused by warming temperatures and ocean acidification. Public VR experiences include the free *Chasing Coral* by Exposure Labs and The Ocean Agency (2016), David Attenborough's *Great Barrier Reef* preview by Atlantic Productions and full-length film shown at the Natural History Museum in London (2018), and *Sanctuaries 360°* provided online by the National Marine Sanctuaries of the U.S. National Ocean Service (2021). These VR films have an educational purpose, funded respectively by a conservation agency, museum, and marine sanctuary, with *Sanctuaries 360°* even including lesson plans to complement the VR experience in the classroom (National Marine Sanctuaries, 2021). They are also a form of climate activism that uses the technology of VR—combined with widely available smartphones, tablets, and computer gaming systems—to bring what scientists recognize as the plight of the ocean's coral reefs and their bleaching to the public.

This modality works at providing an "immersive" experience, not only through the physical 360° visual restriction of the user's gaze, but also psychologically through the reproduction of the underwater environment's scene, in which the viewer is paradoxically situated in a location where they would not usually be able to breathe, and gravity would be suspended by the buoyancy of saltwater. The VR reproduction allows the user to view the coral reef without feeling the coolness of the water, its waves and motion, the sting of the water's salinity, and the sharp coral edges as they brush by them. But for most people, the virtual rendition is as close as they will ever get to experiencing a coral reef. The reproduction of the coral reef through VR attempts to make this experience and the knowledge that it would generate more widely available to the public, regardless of socioeconomic status or physical abilities.

However, VR, while giving the impression of offering freedom from normal constraints, is highly scripted. This technology dislocates traditional embodied understanding for a simulation that cannot fully reproduce or replace physical experience. In addition, users are often aware that their senses are being manipulated, no matter how smooth an experience the technology might be able to give. This effect likely diminishes the believability of scientific content that is shown by the VR experience and could lead the user to feel as if he or she needs to "suspend disbelief," resulting in the opposite effect of what was intended. As Leighton Evans and Mark Rzeszewski note, "the conflation of immersion and embodiment minimises the attention that can be paid to the creation of the sense of place and placehood in VR" (Evans and Rzeszewski, 2020). They suggest a re-evaluation of the "I – technology – world" schema through post-phenomenology, or mediation theory to instead evaluate success in VR immersion through embodiment and development of a virtual presence

(Evans and Rzeszewski, 2020, p. 6), an argument they further refine in chapter 8 of this edited collection. However, including embodiment and subjective experience in scientific knowledge-making tends to disrupt the ethos that is developed through the scientific process. Scientific knowledge is prized because it is objectively derived through deductive reasoning with the positivist ideology that there is one truth that can be discovered through observation by the common faculty of human senses. However, during VR immersion, the subjective bodily experience of the user is manipulated, split between virtual and physical realities, causing two perceived realities at one time, which is illogical according to the tradition of science. Considering this problem, how can VR live up to its potential to give students an immersive, engaging experience while addressing possible doubts about veracity? In addition to promoting environmental empathy, how can VR be used to play a role in addressing the many problems that arise from climate change?

To examine these questions, I include a review of the literature on VR technology that investigates its current and potential use in educational settings to improve learning and practical outcomes. While problems with access to education have always existed as a form of gatekeeping in society, modes of information access have increased exponentially with the explosion of new communication technologies that have proliferated in the past half a century. Here, I include a direct look at VR's affordances and evaluate its usability in the context of the classroom and scientific investigation for climate systems modelling. Climate systems models are "mathematical representations of the Earth that help us predict changes in the climate... [they] provide simplified versions of the real world that can be used to test our understanding of how the climate system will respond to changes in conditions," allowing us to forecast future climate conditions and adapt to obtain better survival rates (IPCC 2013). My findings point toward several major differences between the physical experience and the VR simulations that could be bridged with improved functionality of the VR devices and/or content of the VR programs. The VR simulations lacked the element of unpredictability and danger, the tactile sensory input, and the opportunity for visual remediation that the real-life experience offered. These limitations effectively turn the VR devices into slightly more interactive, complex 3D film viewers that are not much more effective at promoting learning than watching a TV program.

Despite these limitations, I argue here that VR has immense potential as a tool for climate systems modelling in the classroom and for scientists, but that the simulations created by programmers for the purposes of climate scientist training and recommendations for policymaking need to rival those used in the medical professions to teach surgery on the human body. That is not to say that VR created to promote empathy for environmental distress does not have a role in urging the public and even students and scientists toward action and support of policies toward environmental sustainability, but that VR can have a more practical role in solving the climate crisis.

The unique properties of VR in education about environment

Much recent research on VR and environmental communication involves measuring the ability of VR to elicit empathy in users. At the Stanford Virtual Human Interaction Lab's environmental education lab Hydrous, participants in the Stanford Ocean Experience (SOE) use the HTC Vive VR system see "the ocean absorb carbon dioxide (CO2) molecules, a coral reef degrade, and marine life disappear as the ocean acidifies" (2020). Principal investigator Anna Queiroz describes the purpose of her research as finding ways to promote environmental literacy, where the individual's actions are directly linked to their effects on the environment. Her team at Stanford is comparing the affordances of different VR programs for educational purposes and is even attempting to measure the effects of "double immersion," in which VR users are immersed in water to have the embodied experience of swimming while viewing a virtual environment through a waterproof headset (2020). Results from one of the lab's studies, based on survey data from teachers, claim that "lack of science literacy, unprepared education field, complex and invisible nature of OA [ocean acidification] and lack of personal connection with the ocean" were the main barriers to education about the sources and dangers of ocean acidification (Fauville et al., 2020, p. 1).

These factors, as stated by the investigators, put most of the blame for public apathy and/or misunderstanding of coral reef bleaching on the individual and their supposed "lack" of literacy, education, ability to understand complex information, empowerment, and empathy. It is not surprising, then, that these researchers claim that VR can solve this "lack" by educating the public, improving their feelings of empowerment to make change, and increasing their empathy for the ocean environment. However, this view of how VR should be used to engage the public with environmental issues through pathos and the transfer model of education (where students are given set information that is transferred from teacher to student without the co-construction of knowledge) is not as productive as it could be if it were geared more toward collaborative learning. In the transfer model, students do not have the agency to create new knowledge from the information they are given, and their unique understanding of climate issues remains locked away by this method of pedagogy. This lack of co-construction not only limits the promotion of free inquiry in science but also students' feelings of efficacy and intellectual self-worth. When users realize that they are limited by the program created by these scholars, real, collaborative knowledge-making is not possible in this one-way virtual environment. While there is a temporary benefit to the VR modality from its novelty in terms of student interest, once accustomed to the technology, students look for ways that they can manipulate and use the content at hand.

In addition to the work at Stanford, other studies have also found that use of VR applications in educational settings can promote better attitudes toward learning (Liu and Tsai, 2013), enhanced decision-making abilities, increased opportunities for learning, and a safe setting in which mistakes do not result in catastrophic failures. One of the most important benefits of VR technology that is not as easily offered by other modalities is its inclusion of physical movement in the learning process. Studies done by Richard Lamb on the effects of VR on students' scientific writing have even shown that the modality increases the physical human hemodynamic response in neural activity when compared to responses to purely written material (2020). The positive effects of bodily movement on the learning process, when combined with the virtual environment, is something that needs to be studied further, and could result in breakthrough educational progress. In the traditional classroom, students are expected to sit in rows in their seats with minimal movement, but this physical arrangement is perhaps not the best way to educate young bodies that need exercise to stimulate their minds. Even Aristotle took his pupils on long walks while discussing philosophy, and the school was called the "Peripatetics" because of their constant movement.

While research on VR for science education does show benefits, I argue that using VR for science communication can also come at a possible cost. The paramount social status that scientists have worked to develop is founded on trust in their research and the accurate representation of findings through scientific and technical documentation. VR not only prioritizes embodied understanding over rational dissemination of information through traditional means (the genres of peer-reviewed articles, conference proceedings, technical reports), but also offers a simulacrum of reality that can never accurately replace physical experience. Here, it is useful to review what gives scientific and technical documentation *ethos*:

- the certitude that the research is peer reviewed and has its basis in a developed field of knowledge (shown by the known viability of the place of publication, i.e., the journal, book series, publisher, author and institutional affiliations);
- the organisational tropes of genre are followed in the content;
- and finally, individual or personal experience is removed from serious consideration, which is critical to the perceived objectivity of the information.

This is what gives scientific and technical documentation both ethos and separates it from subjective public knowledge. When attempting to accommodate science for public understanding, whether through popular science magazine articles or the newer technology of VR, there is a potentially dangerous process of simplification that attempts to gloss over the incommensurability of scientific research and personal experience (Fahnestock, 1986). Drawing on *pathos* or emotion from users to show the seriousness of the climate crisis, while powerful, will likely not lead to the development of solutions. Placing the blame for climate change on the individual in the form of shaming and guilt might lead to changes in behaviour at the individual level, but even if every person on earth decided to conserve paper tomorrow, it would not change our societal-level trajectory toward climate disaster. However, if that accommodation takes place in a VR system in which the complexity of scientific information is maintained while embodiment and personal experience are still valued, the only accommodation made in the dissemination to the public is that of a change in mode (from written text to the virtual sphere). On the other hand, accommodations that would need to be made on the side of scientists to allow embodied information into the equation would be more disruptive to their ethos in relation to their position as the gatekeepers of knowledge. This impediment can be overcome, however, by comparing climate systems modelling with VR to the models created by medical professionals in the training of doctors.

The potential for VR climate systems modelling through Gaia theory

If VR is used to promote greater complex systems intelligence through (virtual) physical interaction and serious games—a model already used by the healthcare industry to train medical students about treating the systems of the human body—it could be a major contribution to climate science. A review of the literature on medical training applications describes VR as having a high potential for improving learning and accuracy in medical treatment contexts, most widely used first in the context of laparoscopic surgery (Samadbeik et al., 2018). This type of surgery is done by inserting surgical tools inside the abdominal cavity to perform a variety of operations using an internal body camera, so it has lent itself readily to training in a virtual environment with a simulated patient (Virtamed, 2022). Scientists are currently able to accurately simulate climate feedback systems and interactions and forecast changes over time – if we are serious about wanting to find practical solutions to the threat of global warming, we must create serious game simulations in which students and citizen scientists can manipulate climate interactions in a rich virtual environment such as the surgical training VR program. Envisioning and replicating such interactions in the real world is simply not possible and would be dangerous; therefore, a potentially highly valuable future for using VR in promoting environmental sustainability would be to model or prototype human-environment interactions on a systems level.

Here, I will explain what a systems approach (Latour, 2018; Rai and Druschke, 2018) can bring to the development of environmental and ecological pedagogy through the medium of VR technology. Scholars studying technical communication are increasingly pointing toward embodied cognition as a prerequisite for human engagement with, and understanding of, environment (Varela et al., 2016; Stinson and Le Rouge, 2022). Rhetoricians have previously investigated the importance of embodiment in communication by referring to cognitive psychology and perception through the senses (Varela et al., 2016), gestural modality (Sauer, 2003), and the importance of physical "place" (Rai and Druschke, 2018) in knowledge-making. Promoting an embodied understanding of environment through the physically immersive modality of VR technology is a logical next step toward this goal. This is a pragmatic description of systems-based learning that is possible through the development of VR programs for environmental interventions based on current medical applications of VR educational simulations.

Fountain provides a foundation for discussing the rhetorical context of the gross anatomy lab in his discussion of ethnographic research done in the medical classroom (Fountain, 2014). He describes a process of increasing student understanding through their physical relationship with cadavers from pictures of the human body to physical dissection and autopsy and scientific reorganisation of the human form. He claims that previous research "demonstrate[s] that much of the labour in technical communication is distributed across an array of bodies, objects, and even locations" that result in a "trained vision" (Fountain, 2014, pp. 145-146). Since Fountain's study was conducted, there has been an explosion of VR applications in medical training that supplement the gross anatomy lab and give students experience conducting medical procedures such as laparoscopic surgery in a low-stakes simulation of patient-doctor interaction (Samadbeik et al., 2018). Such educational programs, while costly to develop and implement, have been shown to improve learning speed and accuracy in recognizing anatomical positions and result in better surgical outcomes (Samadbeik et al., 2018, p. 126). The VR program cannot replace real experience in the operating room; however, it can prepare students to take on the role of surgeon by increasing awareness of the physical context they will experience when they are given the opportunity to participate in a real-life operation.

To make the leap from VR simulations used in the training of medical students operating on the human body to the training of environmental students operating on the Earth's body, it helps to operationalize Gaia theory in the concrete manifestation of ecological systems. Here, we are looking at Gaia theory as a practical construct to support a systems-based investigation of climate change through the technology of VR. Gaia theory was developed by James Lovelock in 1965 and is the "hypothesis that living organisms regulate the environment in their own interest" (Lovelock, 2003). The theory received much criticism because not only did it suggest that the Earth could only be truly understood by looking at its functions holistically with a systems view, it also did away with the division of scientific fields that is the basis of much of Western knowledge. If biological functions are constantly attuned to changes in geological and atmospheric functions, then what use is there in having separate fields of study? In 1981, Lovelock "composed a model of dark- and light-coloured plants that competed for growth on a planet in progressively increasing sunlight ... not to make a blueprint for the Earth, but a model to show that Gaia is consistent with natural selection," which he called "Daisyworld" (Lovelock, 2003, p. 769). He later found evidence that this model was a serviceable representation of the feedback loop that existed between organic and inorganic matter on Earth, which is altered by temperature. Climatologists now incorporate responsive biological feedback into their models that forecast future changes in climate that increases their accuracy.

Daisyworld is a Darwinian model of planetary self-regulation, "a parable, invented to demonstrate that such self-regulation can occur without teleology (conscious foresight or planning on the part of unconscious organisms), from a tight feedback coupling between life and its environment (Watson and Lovelock, 1983)" (Lovelock, 2000, p. 109). In Daisyworld, black and white daisies compete for sunlight and growth, and in turn influence the surface temperature of the planet in different ways because of their varying albedos (amount of light reflected or absorbed). However, there are constraints to how quickly the daisies can adapt to temperature changes, and whether white or black daisies dominate on the planet is constantly in flux. Understanding these systems constraints is key to developing accurate models of climate change in VR that can be manipulated by students to gain a greater understanding of how the Earth functions as Gaia. Lovelock argues that "there are bounds on the range of environmental conditions that can be adapted to, and as these are approached environmental conditions limit growth rates and often demand adaptations to survive" (Lovelock, 2000, p. 114). Conditions at the acceptable limits of habitation, where the tipping point of adaptation exists, hold the information necessary to promote both human-controlled adaptation in biological systems and geoengineered environments toward planetary sustainability.

To view the human body as one holistic organism is itself revolutionary because medical science has been divided into all the separate organ systems that make up the bodily functions: the circulatory, endocrine, immune, muscular, skeletal, nervous, and digestive systems, and so on. Studying a separate system such as the skeleton is only useful in the full context that bones attach to muscles that are both fed by veins and operated by nerves. In addition, it is not only the different physical systems of the body that must be seen as interacting with each other, but also the mind with the body. Robert Reiner and Matthias Harders make the distinction in the medical use of VR between *presence*, or "the subjective experience of participants in a virtual environment" and *immersion*, or "the capability of a system to display an artificially generated environment in a way that it comes close to the real experience" (Reiner and Harders, 2012, p. 2). VR promises high fidelity to *immersion* for students of the human body and interventions that can solve problems with its proper function. The twin goal of *presence* acknowledges how individual, embodied experience matters in the learning context. This is the basis for a holistic view of the human body's operation and how it tracks with VR technology.

As discussed earlier, medical training with VR has been shown to improve both student educational and patient outcomes (Samadbeik, 2018). Companies such as Osso VR and ImmersiveTouch provide VR medical training and assessment programs to schools that want to improve student success rates (Osso VR, 2021; ImmersiveTouch, 2021). However, apart from many articles discussing the features and functions of such training programs that replicate a certain procedure and evaluate their ability to educate, there has been little written about how medical VR applications also operate at a systems level, where the human body is holistically studied. An operation on one bodily function, such as the pumping mechanism of the heart muscle, will have immediate effects on lung capacity and neurological firing in the brain. Both the body and mind play a role in repair and recovery. Such effects are effectively simulated by artificial intelligence algorithms so that medical practitioners have a fuller understanding of how a person will react to different types of interventions.

Yuan suggests a holistic medicine and biology model that draws on Eastern healing practices (Yuan, 2017 and 2019). He argues that "studies in complexity science have shown that living organisms are inherently self-organizing, self-adaptive, and self-regulating" and the move from reductionist science to a holistic view will result in doctors offering more effective diagnoses and treatments (Yuan, 2019, p. 23). The move toward holistic medicine also means developing methods for individualized treatment based on a person's unique physical makeup and emotional state. This systems intelligence is more practical in the provision of modern medical treatment with the resources that can be developed with the modelling capacity of AI and scaffolded with VR technology. According to Yuan, "In traditional scientific fields such as physics, astronomy, ecology and biology, new knowledge structures based on holistic methods are emerging. Through a complex object modelling process, state variables and measurable output selection can be built upon the foundation of the analytical scientific method" (Yuan, 2017). Thus, it is theoretically possible to take a holistic approach to climate systems modelling that reflects holistic medical practices.

The technical and moral case for embodied systems modelling in VR

Modelling climate intervention VR programs after medical training programs can give geophysics scholars a useful tool with which the Earth and its systems can be evaluated holistically. Even if the metaphor of the Earth as Gaia, with processes analogous to the workings of the human body, does not match up perfectly to the reality of the Earth's geophysical mechanics, it still gives scholars the opportunity to explore how this complex system operates and to find its limits. If we think of the Earth as a system, much as we think of a human body, then it is possible to measure and evaluate how interventions will affect the Earth over time. The boundaries of the systems common to our conception of the human body and the Earth include:

- The human body and Earth are both born at one point in time and will die at a certain time.
- They have a state of relative health or illness caused by internal and external factors.
- An intervention in one system of the organism will affect all systems.
- They self-regulate, or adapt, for survival within limits.
- The laws of physics, such as conservation of energy, gravity, apply to both.

We call different eras in planetary history "ages" and even attribute "growing pains" to the Earth's moving crust and volcanic eruptions. There are scientific estimations of when the Earth was formed (4.5 billion years ago), when life began on Earth (3.7 billion years ago), and when it will become uninhabitable (varying from 300 million years to several billion years from now). But it is certain that the Earth will not sustain life forever, and this fact makes statistical modelling of the planet's progression toward this end mathematically possible. Intervening factors toward the dissolution or entropy of the planet include changes in the chemical and biological makeup of the atmosphere, solar energy and its transmission, and gravitational changes in the rotation of the Earth, all of which can be affected by the actions of humans on the planet. However, as has been posited by Lenton and Lovelock, these systems also interact to balance out and self-regulate toward the preservation (or conservation) of energy in a Daisyworld-like continuum (Lenton and Lovelock, 2000). If we truly seek to intervene in the operations of the planet toward the aims of human self-preservation, then we need to come up with much better, more comprehensive models of these complex interactions that combine the speed and precision of calculations conducted by artificial intelligence and a human interface that operates just as fluidly between body and mind as VR.

I have delineated the practical basis for viewing the Earth in terms of the human body, providing a technical basis for the creation of VR programs that might extend similar functions of operationability from human surgery to treatments meant for Gaia toward improving planetary environmental sustainability. However, there is also a moral imperative to acknowledge that the Earth can experience relative states of health or disease that is gaining ground in legal spheres. If we extend the analogy to give Gaia personhood, then it is necessary to give Gaia the same moral rights that we normally accede to humans, as delineated in the American Declaration of Independence: the right to "life, liberty and the pursuit of happiness." Movements to grant legal personhood to geographical entities have been increasing and gaining ground lately, with legislative protections sought for rivers in the United States, Colombia, India, New Zealand, and Australia (Clark et al., 2020). Cristy Clark and colleagues point out that "Nature, the environment, and even single complex ecosystems, are seldom easily quantifiable as bounded entities with geographically clear borders" (Clark et al., 2020, p. 787). However, as opposed to riverine systems, the Earth has well-defined boundaries that allow a clear-cut pathway toward legal protections based on ecological jurisprudence. A legal definition of Gaia granting personhood and rights further supports the notion of orienting VR climate systems modelling toward a holistic body-planet operational structure that also promotes empathy for the climate crisis as embodied in Gaia.

So, in relation to both technical form and social practice, it is technically possible and morally imperative to build climate systems models in VR that give geophysicists and bioengineers a workable tool that not only forecasts weather on the scale of days or weeks, but also can forecast how human interventions in climate will play out in multiple systems over decades and centuries. If we take Lovelock's Daisyworld model as an example of a simplified planetary "parable" developed from Gaia theory and build upon it with multiple interacting systems variables that have limits as defined above, mathematical calculations of probable future trajectories for planetary stability/instability are possible that can be coded for visual/haptic/audio representation in a VR system. Such rendering of environment in VR has, as discussed earlier, already been done in less interactive form in the context of coral reef bleaching events, and scholars are working on improving interactivity with projects such as the LANDIS-II Wisconsin forest climate scenario VR experience (Huang et al., 2019). By applying the technological functionality of medical training programs in VR to climate intervention models, interaction between human and planet can be processed both quickly and efficiently while sustaining cognitive understanding and empathy for human users.

For example, the VirtaMed LaparoS VR simulation training system for laparoscopic surgery includes a plastic model of a human torso on a medical cart with wheels and a computer screen. The same tools that would be used in the operating room are provided, such as blunt dissection and electrosurgical instruments, guided by a small laparoscopic camera (VirtaMed, 2022). It even gives a similar physical sensation to lifting and moving organs and tissues and provides medical information on-screen during the simulated procedure. A VR climate simulation model could include both a physical model of the planet and a virtual rendition on screen that would reflect the effects of different physical interventions such as levels of CO2 and sunlight, increased or decreased plant cover (as in Daisyworld), and varying ocean levels and atmospheric patterns. These simulations would need to also consider how planetary inventions would affect habitats for humans and other living beings on earth. This is a very basic wireframe for a potential VR program that could simulate planetary interventions in climate, based on medical training VR, Gaia theory, and the Daisyworld model, which needs more development in the future.

My argument has been that VR as currently conceived in the context of environmental communication spends too much time and resources on promoting empathy, and while chapter 7 in this collection shows the benefits of environmental empathy, it also has limits. Consequently, current climate system models created by scientists tend not to enrol VR for the purpose of promoting human – environment virtual interaction and operationality toward finding solutions to the climate crisis. We must move forward with collaborative efforts to solve environmental sustainability issues by allowing an interactive relationship between the educational VR user and climate systems models. In return, climate systems scientists need to take up the technology of VR to more accurately model Gaia in terms of the intersecting systems of the human body, even if traditional models of dissemination that uphold the ethos of scientific information (through purported objectivity) are disrupted by the uptake of (more subjective) VR and embodiment in geophysics. This proposal brings an ethics of care (discussed in more detail in chapter 7) to the practice of climate systems modelling vis-a-vis the named personhood of Gaia, which alone promotes empathy and greater understanding of the planet's needs. Lovelock did not have the technology of VR to make his Daisyworld a reality, but we now have access to the tools needed to create such a virtual world. It will take an ideological shift in how this technology is viewed as a pedagogical and scientific tool, matched with a holistic view of the Earth's systems, to make such models realistic and complex enough to allow for successful human-environment interventions toward sustainability.

References

Atlantic Productions. (2018) David Attenborough's Great Barrier Reef. *VR dive.* Available at: https://attenboroughsreef.com/vr_dive.php (Accessed: 23 June 2021).

Bailenson, J. (2016) Stanford Virtual Human Interaction Lab. *Can VR help create empathy around climate change?* TED Talk. Available at: https://vhil.stanford.edu/video/2016/empathy-around-climate-change/#more-11243 (Accessed: 23 June 2021).

Clark, C., Emmanouil, N., Page, J., and Pelizzon, A. (2020) 'Can you hear the rivers sing? Legal personhood, ontology, and the nitty-gritty of governance.' *Ecology Law Quarterly.* Available at: doi.org/10.15779/Z388S4JP7M.

Evans, L. and M.L. Rzeszewski. (2020) *Hermeneutic relations in VR: Immersion, embodiment, presence and HCI in VR gaming.* Lecture notes in computer science. Available at: https://cronfa.swan.ac.uk/Record/cronfa54686.

Exposure Labs. (2016) *Chasing coral: The VR experience.* Available at: https://www.chasingcoral.com/vr (Accessed: 23 June 2021).

Fahnestock, J. (1986) 'Accommodating science: The rhetorical life of scientific facts,' *Written Communication.* Available at: doi.org/10.1177%2F0741088386003003001.

Fauville, G., Queiroz, A. C.M., Hambrick, L., Brown, B.A., and Bailenson, J.N. (2020) 'Participatory research on using VR to teach ocean acidification: A study in the marine education community,' *Environmental Educational Research.* Available at: doi.org/10.1080/13504622.2020.1803797.

Fountain, K. (2014) *Rhetoric in the flesh: Trained vision, technical expertise, and the gross anatomy lab.* New York: ATTW/Routledge.

Huang, J., Lucash, M., Scheller, R., and Klippel, A. (2019). *Visualizing ecological data in VR.* 2019 IEEE Conference on VR and 3D User Interfaces. 23-27 March, Osaka, Japan.

ImmersiveTouch. (2021) *ImmersiveTouch: Revolutionizing digital surgery.* Available at: https://www.immersivetouch.com (Accessed: 14 July 2021).

IPCC. (2013). *Climate change 2013: The physical science basis, frequently asked questions.* Available at: http://www.climatechange2013.org (Accessed: 3 January 2022).

Latour, B. (2018). *Down to earth: Politics in the new climatic regime.* London: Polity.

Lenton, T., & Lovelock, J. (2000). 'Daisyworld is Darwinian: Constraints on adaptation are important for planetary self-regulation.' *Journal of Theoretical Biology.* 206, pp.109-114.

Liu, P.E. and Tsai, M. (2013) 'Using augmented-reality-based mobile learning material in EFL English composition: An exploratory case study.' *British Journal of Educational Technology.* 44(1), Available at: doi.org/10.1111/j.1467-8535.2012.01302.x.

Lovelock, J. (2003) 'The living Earth.' *Nature.* 426, pp.769-770.

National Marine Sanctuaries. (2021). *Sanctuaries 360°: Explore the blue.* Available at: https://sanctuaries.noaa.gov/vr (Accessed: 23 June 2021).

Osso VR. (2021) *Osso VR: The leading, VR training & assessment platform.* Available at: https://ossovr.com (Accessed: 14 July 2021).

Rai, C., and Druschke, C. (eds.) (2018) *Field rhetoric: Ethnography, ecology, and engagement in the places of persuasion.* Tuscaloosa: University of Alabama Press.

Reiner, R., & Harders, M. (2012) *VR in medicine.* New York: Springer.

Saker, M. and Frith, J. (2019) 'From hybrid space to dislocated space: Mobile VR and a third stage of mobile media theory,' *New Media & Society,* 21(1), pp.214-228. Available at: doi.org/10.1177/1461444818792407.

Saker, M. and Frith, J. (2020) 'Coextensive space: VR and the developing relationship between the body, the digital and physical space,' *Media, Culture & Society,* 42(7–8), pp.1427–1442. Available at: doi.org/10.1177/016344372093 2498.

Samadbeik, M., Yaaghobi, D., Bastani, P., Abhari, S., Rezaee, R., and Garavand, A. (2018) 'The applications of VR technology in medical groups teaching,' *Journal of Advances in Medical Education & Professionalism,* 6(3), pp.123–129.

Shin, D. (2017) 'Empathy and embodied experience in virtual environment: To what extent can VR stimulate empathy and embodied experience?' *Computers in Human Behavior,* 78, pp.64-73. Available at: doi.org/10.1016/j.chb.2017.09.012.

Stanford Virtual Human Interaction Lab. (2020) Available at: https://vhil.stanford.edu (Accessed 23 June 2021).

Stinson, S., and Le Rouge, M. (eds.) (2022) *Embodied environmental risk: Problems and solutions toward social sustainability.* ATTW/Routledge.

Varela, F.J., Thompson, E., and Rosch, E. (eds.) (2016) *The embodied mind: Cognitive science and human experience* (Rev. ed.). Boston: MIT Press.

Virtamed. (2022) Medical training simulators. Available at: https://www.virtamed.com/en/medical-training-simulators (Accessed: 3 January 2022).

Yuan, B. (2017) 'Holistic medicine: A new medical system based on body constitution and functional status,' *Journal of Traditional Chinese Medical Sciences,* 4, pp.3-9.

Yuan, B. (2019) 'Toward holistic medicine and holistic biology: life sciences after precision medicine and systems biology,' *Frontiers in Life Science,* 12(1), pp.14-26, DOI:10.1080/21553769.2019.1682064.

Chapter 7
Designing with care ethics in nature-based VR

Kailan Sindelar
University of North Florida

Abstract

Nature-based VR is a promising tool for educational and therapeutic purposes, which prompts consideration for how wildlife and wilderness are represented and what kinds of connections are cultivated between users and representations of nature. Care ethics provides a way to prioritize connections between users and other forms of nature. In this chapter, I look at how nature is formed in three popular VR apps and present ways to incorporate care ethics into future work. I suggest that designers include human relationships with more-than-human nature, let materials guide their design, and work towards representing the complexities of animal lives.

Keywords: nature, framing, care, ethics

***

Introduction

Nature Treks VR features eight different scenes where the user can walk around and explore. These scenes do not replicate specific places at a point in time. Instead, Nature Treks is more like a moving portrait where users can navigate around a circular space and look around in 360 degrees. Within the circle, there is abundant flora and fauna related to the scene. In the Green Meadows scene, many rabbits, birds, and deer pass by. In the distance, there are mountains and more wildlife moving around. The sounds match each space and reflect what the user is near, such as a brook or a bird. The scenes pass by peacefully, and there isn't anything for the user to do but move around or simply stay still. A similar, slow progression plays out in each of the scenes that vary between snowy woods, deep oceans, tropical beaches, and other depictions of wilderness.

Nature Treks VR is an example of a still-emerging genre of nature-based VR. Research has shown that some nature-based VR experiences can decrease perceived and remembered pain (Tanja-Dijkstra et al., 2018) and increase stress recovery (Annerstedt et al., 2013). In particular, *Nature Treks VR* was shown to help persons with trauma and stress-related disorders better than similar non-VR experiences for intervention and general well-being (Frewen et al., 2020). These nature-based VR apps mimic the feelings and effects of being in actual nature (Breves and Heber, 2020; White et al., 2018), but these experiences are more cultivated than going into actual outdoor areas. Though apps like *Nature Treks VR* can be a useful tool in managing anxiety and stress, they also present limited, albeit useful, views of the wilderness. These limits are important and should be considered by researchers who examine how wilderness is represented through VR and what kinds of connections are cultivated between users and representations of nature.

Nature Treks VR is far from the only nature-based VR experience. Other VR apps such as *Ocean Rift* and *National Geographic Explore VR* use computer-generated images (CGI) to simulate natural environments. The environmental focus in *Ocean Rift* and *National Geographic Explore VR* leverages some of the unique affordances of VR, which are particularly well-suited to environmental communication. Environmental crises pose considerable communication challenges, including incorporating first-hand experience, experimenting in specific environments, providing visuals for typically imperceivable phenomenon, and bridging the spatial and temporal distance between people and places (Fauville, Bailenson, and Queiroz, 2020a). VR is well suited for overcoming all of these challenges, and research has shown it is able to increase user knowledge and an interest in learning (Markowitz et al., 2018). Nature-based VR experiences can also increase pro-environmental behaviour (Ahn et al., 2015; Ahn, Bailenson and Park, 2014). The ability of nature-based VR to increase beneficial behaviour for environments and mental health for its users makes these experiences highly valuable.

Despite the value of nature-based VR, the ethical consideration for VR design has largely ignored representations of the environment. Some studies have discussed framing through narrative (Ahn et al., 2015) and the importance of matching natural settings with corresponding audio (Annerstedt et al., 2013), but a critical discussion of how environments are selectively captured through video or created through CGI needs to be developed. While VR has been discussed as a media form without a frame (Riggs, 2019), the designers of VR experiences still frame issues by choosing what to represent and how to represent it. The wider access to VR combined with the benefits and uses of nature-based VR prompt further consideration for how environments, and those who inhabit them, are represented. One of the core contributions of this chapter is showing how *care ethics*—explained in more detail later—can be a useful framework through which to consider these representations in VR design.

I argue that representations of nature are framed in VR and that care ethics can help designers create content that prioritizes connections between users and other forms of nature. Considering how people connect to natural environments in VR is especially important because this media form can make users feel like they are in different places with different perspectives. Care ethics prioritizes connection, which includes interaction between humans and nonhumans. I also take a new materialist perspective, arguing that designers need to think critically about what environments and animals in them are expressing. In the following section, I will discuss some notions of nature and how care ethics can apply to nature-based VR. I will then look at the framing of *Nature Treks VR*, *Ocean Rift*, and *National Geographic Explore VR*. I conclude this chapter by detailing ways to integrate care ethics into nature-based VR with attention to framing.

Care ethics for nature-based VR

The nature-based VR apps that I analyse here (and others) could be organized into two large groups, those for human benefit (such as mental health) and those for environmental benefit (such as promoting pro-environmental behaviour). While these two large groups may seem impractically reductionist, this organisation is tempting for designing the experience of the app. Should distressing information about the climate crisis be included in apps that are designed to reduce anxiety and promote mental wellness? These different aims require different information and experiences. However, these two groups *are* impractical. The lines between users, what is typically referred to as materials, and what is typically referred to as nature is far too permeable to allow for such exclusionary groupings. Instead of presenting a framework for when and how people should engage with (other) forms of nature, I propose prioritizing an ethics of care to guide design choices.

Since Carol Gilligan (1993) proposed an ethics of care, there have been influential critiques and contributions, but it is still a theory that prioritizes connection and care over rules for ethics. Gilligan's theory has been very popular with feminist moral theorists across a wide array of topics. Care ethics does not draw impermeable lines between humans, non-humans, and politics. Instead, care ethics prompts alterbiopolitics, disrupting clear distinctions between different beings and rules between them. Maria Puig de la Bellacasa explains this idea, writing:

> "Care troubles and opens questions here [obligations, contracts, and promises]. Connecting the practices of permaculture ethics as everyday ecological doings with a feminist notion of care displaces biopolitical moralities, allowing us to envision *alterbiopolitics* as an ethics of collective empowerment that puts caring at the heart of the search of everyday struggles for hope and flourishing of all beings, of *bios* understood as a more than human community" (Puig de la Bellacasa, 2017, p. 22)

Alterbiopolitics of care ethics is especially useful for thinking about how people engage with nature in VR because it recognizes the permeability of boundaries and gets beyond strict ethical and moral rules.

Getting beyond strict rules involves recognizing both the problems with care and the usefulness of "impure" care. Sometimes care can be uncomfortable because of grieving. Other times care can involve exploitive or abusive divisions of labour. Puig de la Bellacasa explains that care ethics is not about the purity of care; she writes that reclaiming care is not about "purging and 'cleaning' a notion; rather, it involves considering purist ambitions—whether these are moral, political, or affective—as the utmost poisonous." (2017, p. 11). Similarly, Alexis Shotwell argues that a purity approach is counterproductive because "it shuts down precisely the field of possibility that might allow us to take better collective action against the destruction of the world in all its strange, delightful, impure frolic" (2016, pp. 8–9). Recognizing the places of "impure" care is important, especially in forms of media that have multiple aims and work with such a nebulous term as "nature." For example, in an app like *Nature Treks VR,* it is okay to think about how relationships with environments are cultivated without making it the primary goal of the experience. Similarly, in an effort to educate users about environments, it is okay to also consider factors outside of the environments discussed. This framework is also a useful approach for VR (and other nature simulations) because there is no direct—what may be considered a pure—experience of nature.

For the purposes of simplicity, I have used the term "nature" for experiences that replicate or represent wilderness and wildlife (also contestable terms). This is a useful term because it groups together certain kinds of VR experiences that have been shown to support pro-environmental behaviour or have therapeutic effects, such as Annerstedt et al.'s (2013) article "Inducing physiological stress recovery with sounds of *nature* in a VR forest." However, I want to call attention to how this popular use of the term does not draw accurate boundaries around "nature," which can create obstacles when attempting to create a connection between humans and nonhumans. Diana Coole (2010) draws from Merleau-Ponty to describe life itself as a fold. This perspective sees all matter as being and recognizes no point where nature ends and the human begins. The folds disclose nature coming into being in different forms. Feminist materialism puts forth several ways to approach this messy co-creation. Stacy Alaimo (2008) introduces the term transcorporeality, recognizing the transformative power of materials to affect and become parts of different bodies. Similarly, Nancy Tuana (2008) introduces the term viscous porosity, which also addresses the permeability of seemingly discrete material bodies. In Tuana's example of viscous porosity, she (un-earnestly) invites readers to burn a piece of plastic and breathe in the air. She calls attention to how the material has the power to become a change to a body even as it becomes a part of that body (as dioxins from burned plastics are endocrine disrupters). After pointing to how bodies are subject to material agency, she discusses how differently gendered bodies have different experiences with the material world. Connecting people to other bodily experiences is something that care ethics and feminist materialism have in common. This is important for considering how people connect to natural environments, especially in media forms like VR that make users feel like they are in different places with different perspectives.

Care ethics offers a framework for nature-based VR that already prioritizes connection. For example, communicating ocean acidification is one topic that both VR researchers and feminist materialist scholars have discussed. Stanford's Virtual Human Interaction Lab led a research study on communicating ocean acidification through VR. In their app, users go through different places, including near a car emitting CO^2 and deep in the ocean. While users are in the ocean, they interact with pteropods shells that are breaking apart because of ocean acidification (an effect of CO^2 absorbed into the ocean). Working with marine educators, they found that empowerment, perspective-taking, and visualisation are three principal ways in which VR may contribute to educating the public about ocean acidification (Fauville et al., 2020b).

Also writing the difficulty of communicating ocean acidification to the public, Stacy Alaimo approaches the topic from another perspective and examines photographs. The photographs that she writes about are pteropod shells melting away in the acidic ocean. What is striking to her is the absence: the absence of a creature and the disappearing shell. She juxtaposes this absence against images of ocean creatures that are aesthetically posed, writing:

> "Within the contemporary digital landscape in which ocean creatures are posed in highly aesthetic ways, by environmental organisations, scientists, and popular media, the shells take up their place in the virtual gallery of aesthetic marine pleasures, haunted by the missing fleshy life. To say that they call us to contemplate our own 'shells'—or a bodily and psychic boundaries—on acid suggests something akin to a psychedelic experience." (Alaimo, 2017, p. 111)

Stanford's VR study and Alaimo's (2017) observations seem to be at opposite ends of a spectrum. The VR experience is noted for its ability to depict something that is inherently difficult to see. Alaimo points to objects that make us aware of absence, not creating a visual for it. However, each example allows for users/viewers to make connections between bodily/material experiences. These bodily/material connections are already important to scholarship on VR as a form of media *and* to scholarship on environmental communication using feminist materialism. Yet, these areas of study rarely converge. Not only does care ethics fit the goals of nature-based VR, but it also creates opportunities to learn from other media that let material effects guide their design. This is an important bridge between otherwise unconnected scholarship, but differences between media forms are still significant here. Alaimo's photographs have obvious frames and the frames in Stanford's experience are more difficult to recognize, a topic addressed in the next section.

Framing in VR

Framing, both literally and figuratively, is an important part of VR design that is often understated or misunderstood by the notion that VR is "frameless." Beyond narrative framing, the selection of audio and visuals inherits the design programs and expertise of other genres, such as gaming, photography, and documentaries. There is no VR experience that replicates the world in a way that even comes close to the depictions in *The Matrix* and other pop-culture books and movies. Instead, VR experiences are often very limited. Apps often break up content into shorter playable sections, which allows users to take breaks without disrupting their experience. VR environments also, like all other media, have boundaries. Those boundaries do not look like traditional frames of other screen media (such as a rectangular box), but they are framing boundaries, nonetheless. Perhaps the hesitation to see these boundaries as frames comes from the typical ability to look outside of frames. A viewer can easily look beyond a TV frame to see their surroundings. The lack of easy opt-out is partly what makes head-mounted displays (HMDs) so powerful. In a study where participants watched nature documentaries through web VR while others watched through an HMD, they found that people using an HMD had a higher level of spatial presence and that users perceived more direct experiences with nature (Breves and Heber, 2020). The lack of being able to look outside of a VR experience negates a traditional kind of frame and has powerful effects. However, the absence of a traditional frame does not mean that there are no other framing devices.

The "nature" in nature-based VR isn't simply decoration; it is central content that designers build (and frame) their space around. This point is important because, as Melissa Bosworth and Lakshmi Sarah (2019) address, designing VR is more spatially focused than other forms of media. They quote Joel Douek of Ecco VR, who claims that VR is unlike film and television because in VR "you can just be in a place, like underwater in the Galapagos" and "there's no need to be a story, it's just a place" (Bosworth and Sarah, 2019, p.198). Users often engage with spaces on their own time. The VR designer cannot know how long the user will take in any one spot or where they place their attention over time. This means that designing VR experiences is more about managing content in space rather than over time. This self-guided pacing and view is a kind of framing in VR. In an interview with another VR creator, Sarah Hill of StoryUp, Bosworth and Sarah quote her saying, "so there is a frame, it's just that it's rotating inside the sphere, and it's your job as a storyteller to ensure they're looking in the appropriate direction" (2019, p. 197). Another industry creator, Max Soloman of Black Dot Films, talks about this challenge and calls guiding the users frame an "induced edit." (2019, p. 197). Any time spent designing a VR experience forces the creator to think about frames. For example, designing an app in Unity (a common game design tool that is used for VR), requires thinking about placing objects in a limited area even as the possibilities expand into a (seemingly) endless XYZ grid. Framing an experience is essential on a conceptual scale as well as in its actualisation. If a user runs off in one direction where there is seemingly nothing to interact with, that is first and foremost, an area that the designer chose to create. It is not a neutral infinite space. Secondly, this would take users away from whatever content the designer did intend for them to engage with.

Examples of framing in nature-based VR

Nature Treks VR, *Ocean Rift*, and *National Geographic Explore VR* share significant qualities. These experiences feature places that users have likely never seen (such as the arctic), cannot immediately access (such as a tranquil meadow with deer and rabbits), and activities that are difficult at best to access (such as diving with sea turtles). All three of these nature-based apps rely completely on CGI and not photographs for their experiences. There are many VR experiences that use videos of wilderness exploration or adventure, and while I do not have room to discuss those here, I do want to acknowledge that they also make choices to include and exclude. This process typically involves choosing where to take the camera, where to stage actions, and how each scene is edited. For example, in a VR experience of diving with sharks, the sharks may be baited to come close, but they still exert relative autonomy in how they move and act. This isn't the case for VR experiences that do not use photographs or videos. The apps I am discussing below are useful examples because they present a range of goals and demonstrate how entire natural environments are designed.

Nature Treks VR is a particularly interesting example of nature-based VR because of the research on its mental health benefits (Frewen et al., 2020), but this app was not created by researchers. It was created by a game developer who has worked on major videogames published by large corporations such as Sega. It is the first app from the company Greener Games, whose aim is "to do more research on possible therapeutic gains using an interactive medium" (*Nature Treks VR*, 2021). The therapeutic aim of Greener Games' products necessarily excludes certain kinds of natural phenomena while including others. Consequently, *Nature Treks VR* provides a particularly curated simulation of nature. Describing the goal of *Nature Treks VR*, the creator John Carline writes:

> "Leisure time is difficult to obtain [...]. This means a high percentage of people miss out on the sounds, shapes and colours of our natural environment. Nature Treks uses technology as an aid to bring some of the healing properties of these into peoples [sic] homes in a convenient and stress free environment." (*Nature Treks VR*, 2021)

Looking at the developer's stated goals and at the VR experience, we can see that attention to colours and shapes are the primary form of engagement in this app.

As I described previously, there aren't many prompts for interaction in *Nature Treks VR*. Users can navigate a circular area, and they can listen to sounds that correspond to their surroundings and look at animals slowly moving around from different vantage points. There are some exceptions where animals move away when users touch them, but each of the scenes in the app is designed for passive engagement as the world moves around the user. While this design may not use much interaction to engage its user, the app engages users in its intended way—by reducing anxiety through depictions of nature. Interaction doesn't frame these scenes, but the scenes do have framing elements. Animals and objects are staged around each scene, prompting users to look in certain directions at certain times. Additionally, each scene has clear boundaries, limiting the user's experiences to central perspectives. These perspectives include looking out at mountains in the distance, deer nearby, and never knowing what lies beyond the immediate area.

National Geographic Explore VR features more influence from actual environments, and it is based upon a creative variation on a rail shooter style design. Rail design is where a user follows a track through a world (visible or not), and they are common in VR experiences. A popular example, and one that is often videotaped for reaction videos, is roller coaster simulations. However, this track design has earlier roots in videogames. Rail shooter design allows players to rotate in their "car" and shoot at objects as they move along the track.

Some popular non-VR examples include parts of Nintendo's *Captain Toad Treasure Tracker*, *Super Mario Sunshine*, and *Pokémon Snap*. This model is a convenient tool because it allows a designer to give the impression of there being a full world to explore—something that the user can spin in any direction to see—without having to design any depth beyond the tracks and interactive elements. It is an especially well-suited tool for VR because it makes use of 360-degree vision without the time-consuming task of designing a whole open world.

This rail design is an extremely temporally bound design (unlike some VR media that is less temporally bound), but the model used in this app still allows for more self-guided time "off" the rails than other, similar designs. *National Geographic Explore VR* uses this rail design creatively by allowing the user some extra control and by using other intermittent elements. When the user explores the Antarctic in their kayak, they can steer it right or left to navigate to different objects while slowly being propelled forward on an invisible rail. This design creates a world that is larger than the user can interact with unless they choose to repeat the experience. This design choice also introduces some of the linearity and time-dependency that other forms of media include. At some points in the rail system, there will always be certain animal activity, such as a sea lion basking, penguins jumping out of the water, and whales coming up for breath. Users of this app play as a photographer for *National Geographic*, so these sites of activity function as interaction points because they are excellent opportunities to take photographs. Both the rails and the activities frame engagement in this app.

National Geographic Explore VR is unique among these three examples for introducing a traditional frame into their VR experience. Capturing images is the main form of interaction throughout the app, and users see their world framed when they look through their camera's viewfinder. Much like taking photographs outside of VR, users must look around their surroundings to choose what to focus on, choose angles, and time action shots. Capturing photographs functions as the exigence for player experience and can be shared on the user's own Facebook feed. This blending of story narrative and user experience is not trivial. Even though different users of this app will travel along the same rail, photographs will reflect personal choices. This app design not only tries to create "induced edits" through animal activity, but it also encourages the user to be aware of how they are framing their experience. This awareness allows for users to think about how they would engage with wildlife and wilderness.

Ocean Rift offers the most open exploration out of these three examples. Unbound to a rail like *National Geographics Explore VR*'s kayak and featuring a more complex boundary than *Nature Treks VR*, this app allows users to weave between obstacles like rock and coral to find new open areas. Boundaries are created by obstacles (such as rocks) as well as invisible lines (such as one preventing the user from surfacing above the water), but they sometimes give way in unexpected areas. These areas reward the user's exploration, featuring new species or a new perspective. For example, in the Deep Ocean scene, the user starts in one open area that has several bioluminescent animals swimming around. If the user swims around this area, they can find shipwreck parts and an audio file trigger that will tell them about bioluminescent sea creatures. Finding and listening to the file is the main method of interaction, and it is framed with sea creatures swimming in large circles around audio files and dramatic backdrops.

Ocean Rift is somewhat like *Nature Treks VR* in how it presents opportunities for engagement. While each scene takes some steps to replicate specific ocean settings (such as a deep ocean with bioluminescent creatures or more shallow depths where sea turtles may be found), animals are stripped of context so that they can easily be viewed together in one scene. *Ocean Rift* also lets users frame their view of wildlife with more control. Users can grab some ocean life and turn the models around so that they can examine them. This design choice is in line with the app's focus on science education, letting users look at animals clinically. However, *Ocean Rift* also rewards curiosity and exploration with new animals. In the Deep Ocean scene, if the user navigates between rocks to a secondary open area, they can find giant spider crabs. While the starting areas and their educational audio files are literally and figuratively central to each scene, the secondary locations add significant depth to each experience. These secondary areas have no additional audio files and could easily be missed, especially in the Deep Ocean scene where much of the experience is very dark. Finding these secondary areas rewards user's exploration with different views. It suggests that exploration and discovery *is* the reward. Though users cannot activate a trigger or grasp an item, discovery is a significant kind of engagement. This discovery and engagement are framed by obstacles in the VR space.

These examples show how boundaries, objects, and actions can frame the experiences of VR. The representations of nature *are* framing devices. None of these apps are fantastical when we look at the pieces they show—rabbits hopping, fish swimming, birds perching nearby—but they are fantastical in their depictions of these things. The degree of wildlife (so many animals circling at once around a person), and the exclusion of balancing forces (predation, limited resources, etc.) are all works of fantasy. Replicating a complete ecology with the innumerable choices within it—where a butterfly lands, where a deer chooses to feed, when a predator decides to prowl—is entirely impractical even if it ever could be possible. However, that doesn't mean that our designs need to be so independent of wildlife influences. These depictions of nature seem to be minimally informed by other kinds of life and their material experiences. These depictions are a missed opportunity to facilitate the connection between people who use this app and the representations of nature that they engage with. In the next section, I address how to use these opportunities by applying care ethics. Currently, these apps create a fantasy of nature that wildlife exists outside of its own contexts and goals. This can be damaging to how people think about nature, reinforcing notions that wildlife exists primarily for the benefit of people.

Framing with care ethics

Nature-based VR is a promising tool for educational and therapeutic purposes. However, like other representations of nature, it influences how people understand and connect to more-than-human nature, so developers must take care in their designs. Care ethics provides a way for designers to think about engagement and framing in a way that prioritizes these connections. Care ethics is not a rule-based system, and it is not limited to what I describe here. This chapter only provides a starting place to begin incorporating care ethics into VR framing. I suggest that designers should contextualize animal behaviour and let materials guide their design, and they should work towards modelling caring human interaction with natural environments. Engaging users in this way can help frame nature as something that includes humans, that has material agency, and that can be engaged with in a caring way. Again, applying care ethics does not mean that designers should take a purist approach where they set out to create the most accurate portrayal of "nature" or to explain *everything* possible. Instead, it prompts designers to think about alterbiopolitics (an ethics of collective empowerment that makes caring central to the struggles for hope and flourishing of *all* beings, of *bios* understood as a more than human community).

Understanding a more-than-human community can be prioritized by designers even if it isn't the central goal of an app. One way to work towards prioritizing the user understanding a more-than-human community is to contextualize wildlife as creatures with their own lives and desires for well-being. We do not see animals fulfilling actions for their own purposes in these nature-based apps. These activities could include gathering food, nurturing each other, or simply enacting any fragment of complex rituals or individual lives that animals have been documented to practice. Though users of this app could never truly (purely) interact with an individual animal, there are no hints that these animals could be individuals. Without even hinting at the individual lives of animals or animal complexity, they are reduced to aesthetic features. While the therapeutic and aesthetic views of nature are appealing and useful, removing all other aspects of wilderness and wildlife reinforces the notion that nature is something that exists for and outside of the human, and ultimately blocks off pathways for care and connection. One way to apply care ethics and get around the problem presented here is to feature context for animal actions and include self-fulfilling activities from these species. This would still need to be filtered for disturbing activities (such as a bear actively hunting), but purity shouldn't be the aim in care ethics. There is room to create apps that benefit people's mental health while facilitating better connections with nature.

More-than-human communities also include material agency. Thus, applying care ethics should include thinking about how materials work rhetorically in an actual environment. The dissolving pteropod shells is an excellent example of this process in both photography and in VR. Instead of materials seemingly existing in a static way, designers of VR apps can think about how materials changing in an environment can frame a situation. For example, in *Ocean Rift*, coral and other physical objects typically frame scenes, but there is little attention paid to these structures. Framing scenes with rocks, coral, and other materials that are carefully informed by the actual setting would allow the material agency of actual places to have effects in the design. When this doesn't happen, whole environments are reduced to simple boundaries of engagement. Instead, these can be places for engagement with material. Places for engagement with material may include the transcorporeal nature of materials and better recognizing the materiality of ourselves. Letting materials inform a design requires some research into an environment, but because most nature-based VR is focused on educational or therapeutic effects, this approach should not be an obstacle. This design can add depth and meaning to an education app, and it can provide more texture for therapeutic nature forms. Besides benefiting the primary goal of whatever the nature-based app may be, this design provides additional opportunities for connection and care with environments.

Finally, and most overtly, applying care ethics can look like modelling caring engagement with natural environments. Engagement, I argue, is both the framing tool and the goal in VR experiences. When VR creators want users to look in certain directions, they want them to mentally engage with one frame of their world over another. Prompts to engage users to create this framing. In the apps discussed above, engagement through animal activity and physical structures leads users to frame their own scene. However, the human behaviour in these apps is unacknowledged at best and disruptive at worst. For example, *National Geographic VR* prompts users to notice how the animals are acting around the user as they move along and explore around a rail design in the arctic, but it does not suggest or model any behaviour other than capturing photographs for the assignment or adventure actions (such as climbing an ice wall or setting up a base camp). Modelling caring actions may mean doing nothing at all. In this example, it could mean keeping the kayak at a distance to give creatures their space. Regardless, modelling any kind of caring behaviour in these apps would acknowledge that the creatures deserve such treatment.

Each of the apps I have discussed here has approached a still nascent media form with the goal of bringing their users closer to forms of nature for some benefit. It is not my intent to devalue their contributions. Instead, I hope that the developers' efforts can help lead to more nature-based apps that build on their work. By building on their work, I mean learning from how VR can be framed and then critically thinking about what kinds of connections we facilitate through those engagements. Again, applying care ethics does not mean that designers should take a purist approach where they set out to create the most accurate portrayal of "nature" or to explain *everything* possible. Instead, it prompts designers to think about alterbiopolitics and how users engage with wildlife. These apps have already shown how VR can play a role in the future of environmental communication; I hope this chapter has shown how care ethics can develop a critical discussion about how people engage with "nature" in virtual environments.

References

Ahn, S. J., Bailenson, J. N., and Park, D. (2014). Short-and long-term effects of embodied experiences in immersive virtual environments on environmental locus of control and behaviour. *Computers in Human Behavior, 39*, pp. 235–245.

Ahn, S. J., Fox, J., Dale, K. R., and Avant, J. A. (2015). Framing virtual experiences: Effects on environmental efficacy and behaviour over time. *Communication Research, 42*(6), pp. 839–863.

Alaimo, S. (2008). Trans-corporeal feminisms and ethical space of nature. In S. Alaimo & S. Hekman (Eds.), *Material feminisms*. Indiana University Press.

Alaimo, S. (2017). Your Shell on Acid. In R. Grusin (Ed.), *Anthropocene feminism*. U of Minnesota Press.

Annerstedt, M., Jönsson, P., Wallergård, M., Johansson, G., Karlson, B., Grahn, P., Hansen, Å. M., & Währborg, P. (2013). Inducing physiological stress recovery with sounds of nature in a VR forest—Results from a pilot study. *Physiology & Behavior, 118*, pp. 240–250.

Bosworth, M., & Sarah, L. (2019). *Crafting Stories for VR*. Routledge.

Breves, P., & Heber, V. (2020). Into the wild: The effects of 360 immersive nature videos on feelings of commitment to the environment. *Environmental Communication, 14*(3), pp. 332–346.

Coole, D. H. (2010). The inertia of matter and the generativity of flesh. In D. H. Coole & S. Frost (Eds.), *New Materialisms: Ontology, Agency, and Politics*. Duke University Press.

Fauville, G., Bailenson, J. N., & Queiroz, A. C. M. (2020a). VR as a promising tool to promote climate change awareness. *Technology and Health*, pp. 91–108.

Fauville, G., Queiroz, A. C., Hambrick, L., Brown, B. A., & Bailenson, J. N. (2020b). Participatory research on using VR to teach ocean acidification: A study in the marine education community. *Environmental Education Research*, pp. 1–25.

Frewen, P., Mistry, D., Zhu, J., Kielt, T., Wekerle, C., Lanius, R. A., & Jetly, R. (2020). Proof of Concept of an Eclectic, Integrative Therapeutic Approach to Mental Health and Well-Being Through VR Technology. *Frontiers in Psychology, 11*,, pp. 858-867. https://doi.org/10.3389/fpsyg.2020.00858

Gilligan, C. (1993). *In a different voice: Psychological theory and women's development.* Harvard University Press.

Markowitz, D. M., Laha, R., Perone, B. P., Pea, R. D., & Bailenson, J. N. (2018). Immersive VR Field Trips Facilitate Learning About Climate Change. *Frontiers in Psychology.* Gale Academic OneFile.

Nature Treks VR—Immerse yourself in a world of relaxation. (2021, May 17). https://naturetreksvr.com/about/ (Accessed January 20 2022)

Puig de la Bellacasa, M. (2017). *Matters of Care: Speculative Ethics in more than Human Worlds* (Vol. 41). U of Minnesota Press.

Riggs, S. (2019). *The End of Storytelling: The Future of Narrative in the Storyplex.* Beat Media Group.

Shotwell, A. (2016). *Against Purity: Living Ethically in Compromised Times.* U of Minnesota Press.

Tanja-Dijkstra, K., Pahl, S., White, M. P., Auvray, M., Stone, R. J., Andrade, J., May, J., Mills, I., & Moles, D. R. (2018). The soothing sea: A virtual coastal walk can reduce experienced and recollected pain. *Environment and Behavior, 50*(6), pp. 599–625.

Tuana, N. (2008). Viscous porosity: Witnessing Katrina. In S. Alaimo & S. Hekman (Eds.), *Material Feminisms.* Indiana University Press.

White, M. P., Yeo, N. L., Vassiljev, P., Lundstedt, R., Wallergård, M., Albin, M., & Lõhmus, M. (2018). A prescription for "nature"—The potential of using virtual nature in therapeutics. *Neuropsychiatric Disease and Treatment.*

Section 3: VR and performance

Chapter 8

Being-there when there is not there: Live performance, social VR, and COVID-19

Leighton Evans

Swansea University

Michal Rzeszewski

Adam Mickiewicz University in Poznan, Poland

Abstract

During the COVID-19 pandemic, lockdowns and social distancing measures put an end to the live music scene almost everywhere across the planet. The lack of live music venues and the ability to create virtual venues for 'live' performance on VR platforms led to the emergence of spaces that could replace live performance. However, the use of VR to remediate live performance raises several questions about the nature of performance and its immediacy, presence for the audience at the performance. This chapter will investigate the possibility of an authentic experience of "being-there" in live performance in VR.

Keywords: live performance, post-phenomenology, technological relations, being-there

This work was supported by the National Science Centre, Poland, grant number 2019/33/B/HS4/00057

***

Introduction

During the COVID-19 pandemic, many governments imposed forced lockdowns and implemented social distancing measures, which at least temporarily put an end to the vast majority of live music across the planet. At the same time, there was also a large increase in digital media usage at home, including in the VR sector of the market (Manavis, 2021). The convergence of the lack of live music venues and the ability to create virtual venues for 'live' performance on VR platforms such as Altspace and Oculus Venues led to the emergence of social VR platforms as a space for replacing live performances. This possibility has been explored from artistic (Berkowitz, 2016; Janer et al., 2016) and technical perspectives (Yakura and Goto, 2020; Geigel, 2017) but gained importance during the pandemic (Chidekel, 2021). Furthermore, the emergence of VR as a viable consumer medium for social activities has offered an opportunity to reconceptualise understandings of immersion, embodiment, and presence in using digital media. VR has always promised to further immersion, which became even more important because a feeling of presence can mitigate the loss of contact with the outside world. Consequently, VR has social and spatial potential to provide space and place for human interactions in times when physical contracts are restricted.

However, the use of VR to remediate live performance leads to several questions about the nature of performance and the authenticity of presence for the audience at the performance. These questions concern the nature of immersion and presence in VR. Many of the discourses and attempts to conceptualise experience in VR conflate these terms rather than understanding each as a state of engagement with a VR environment. This results in a lack of understanding of the importance of design and intentionality in the VR experience with regards to immersion, embodiment and presence. Using a post-phenomenological approach, this chapter differentiates immersion, embodiment and presence as three kinds of relation utilising the *I – technology – world* schema adapted from Ihde (2010). The *I – technology – world* schema dispenses with subject-object distinctions in the use of technological artefacts and positions the device as a co-constituent of understanding the world along with the user of the technology and the world itself. This approach allows for an understanding of these states of engagement as layered and hierarchical rather than instantly emergent on the part of the technology.

The hermeneutic relation between the audience member and the live experience [I (technology – world)] that indicates presence can be understood as a feeling of place or placehood in VR and is intentionally the state often aimed for as optimal in VR experiences. The importance of intentionality as a co-constructor of embodiment and presence is critical in the movement of live performance to VR. To feel "there," the audience must be comported towards the possibility of 'being-there' when there is no "there" for the live performance. This involves both a manner of performance (or intentionality) of the live act and a willingness to both engage with and suspend disbelief in the virtual environment on the part of the audience member. This chapter will investigate the effectiveness of Ç COVID-19 and in the post-pandemic world utilising this framework.

Being-live in Covid-19

Beginning in early Spring 2020 a global pandemic swept through every country, disrupting geographies of every scale and taking its toll on human life. We faced an unprecedented situation where there were artificially induced and controlled large-scale changes - both in their extent and severity - in human spatial behaviour and in the perception of space and place. The net effect of these changes was that people were demobilised on a mass scale. Everyday mobility decreased markedly. We worked from home, we exercised from home, and we swarmed to smartphones and applications to have our favourite cuisines delivered to our doors. Within these changes, some of the economic aspects of COVID-19 become crystal clear. Venues that require people to be present to operate became vacant. Pubs, bars, clubs, and music venues that rely on a live audience for a business model were unable to pivot to a new distribution model for their services. This did not, of course, affect only the businesses that host live performance but also the performers themselves. The arts in general were critically affected by the lack of performance during the pandemic, with live music being an area of the arts and entertainment that ground to a halt completely in March 2020 as lockdowns and social distancing erased the physical and spatial possibilities of performance itself.

During this period, some performers have looked to other ways to provide a live performance experience to an audience. Performing privately online, providing free access to online performance, and using social media as an online means to live performance have all been used extensively. So, unsurprisingly, has VR technology. VR has the advantage of having the potential to provide space and place for human interactions at a time when physical contacts are restricted by the loss of spatial and social dimensions of human experience (Gao et al., 2020). Immersion, embodiment, and feelings of presence in VR can therefore mitigate the lack of contact with the outside world (Slater, 2003). These types of connection have been a long-time promise of VR and its sales pitch – a pitch often focused on replacing the material world in all kinds of human interactions in a dream of a metaverse (Evans 2018). To this end, 2020 saw performances by the likes of John Legend and The Weeknd in Wave VR, Charli XCX in Oculus Venues, and the entirety of Burning Man in Altspace VR during the pandemic.

Even before the pandemic there was a growing interest in moving live performance into the VR (Charron, 2017) which is one of the aspects of a wider phenomenon of deep mediatisation (Couldry and Hepp, 2018). For example, VR offers possibilities of active co-presence with others – an aspect that has been seen as missing from other remote live performances (Tarumi et al., 2017). The flexible and potentially boundless virtual environment can be enhanced with affordances like spatially manipulated sound and special effects connected to users' brainwaves and heartbeats that are impossible in "meatspace" (Atarashi et al., 2018; Horie, Wada and Watanabe, 2017). These affordances provide communication, engagement, and interaction to the live experience (Kharouf et al., 2020). Live performance platforms such as Oculus Venues, Altspace or Wave allow players to choose avatars to represent themselves and to be present in performances and interact with other users. These platforms are available for a wide range of VR headsets and can be also accessed in desktop mode. Some allow for full-body tracking that can be used with appropriate equipment (e.g., Vive Tracker) to bring additional modes of embodied social interaction. While these platforms share the promise of providing an immersive social experience to users, the critical aspect is, for live performance, avatars. Avatars are known to have the potential to help invoke a sense of presence through building a relation between the physical and virtual body (Slater. et al., 2008) and this is even truer for immersive environments where avatars play an important role in embodied cognition (Steed et al., 2016).

The key element that makes VR unique in a spatial context is that it offers a new mode of mediated social interaction, and *immersion* and *presence* are two terms that are frequently used to describe the experience of VR (Evans and Rzeszewski, 2020). While immersion is "simply what the technology delivers from an objective point of view" (Slater, 2003, p.1), presence is contrastingly understood as being "a human reaction to immersion" (p. 2). Immersion, as Evans (2018) argues, is a unique selling point of VR. Engaging with VR has often been considered an immediate gateway to the experience of feeling immersed. In psychology, the term immersion is used to describe a state of mind in which the participants are "completely involved in something while doing the action(s)" (Muhanna, 2015, p.347). Presence is also frequently seen as a kind of product of immersion, where more advanced immersive technology evokes a greater sense of presence (Cumming and Bailenson 2016).

Evans and Rzeszewski (2020) argue that the critical difference between a sense of immersion and a sense of presence in VR is due to the relations among the human user, the technology, and the interface and the world itself – in essence how the human and technology work together to remediate the experience of the world. An experience of presence requires a focal relation, an embodiment relation, and a hermeneutic relation where the meaning of the existential locale itself is constructed through the relation of human and technology towards the world (Evans, 2015, p.7). Presence is, therefore, both a phenomenological state of being and a construction which is contingent on the technology of VR, the interface of VR, the design of the experience, the content of the experience, and the intentionality and mood of the user towards feeling embodied and present in a VR experience. In the context of social VR then, the use of a head-mounted display (HMD) alone is not enough for presence; the VR user must also be oriented towards a sense of embodiment and presence, and the VR environment must be accommodating to this orientation. In the next section, we investigate how live performance has been created through these three layers of relation in VR and how this kind of live performance critically differs from "meatspace" live performance.

Being-live in VR: place and physicality

There has been, since the first wave of VR in the 1980s, an assumption that VR has the technological tools to invoke a sense of place and presence through immersion and embodiment mechanisms. This sense of place is very important in live performance– it is one of the main motivations to attend such events (Brown and Knox, 2017). However, with VR performances, the place itself has been transformed by the digital technologies. This transformation is conspicuous in the mediatisation of live performances, where being-in-place means also being in many other places at the same time through the use of digital media that produce the eventsphere that conjoin distance and proximity and shape the sense of place (Hammelburg, 2020).

Place has always been perceived as natural condition of human existence in the tradition of phenomenological thinking (Heidegger, 1962; Tuan, 1975;) and the possibility of media technologies being responsible for diminishing the sense of place has been the source of considerable debate (Relph, 1976; Augé, 1995; Meyerowitz, 1986). A sense of placenessness for a live performance would be a death knell for the experience of that performance; people would be unable to feel ground and feel a connection that makes them want to dwell there with the performance. However, it is increasingly common that places are created solely through the use of digital media and those practices of "re-placing" becomes embodied in spatial habitus (Halegoua, 2020). The re-emergence of VR occurs, therefore, in a media landscape already rich in social mechanisms that adopted digital technologies as one of the components of everyday spatiality. And the COVID-19 pandemic was likely seen by many people as a reason to incorporate VR into their own spatial practices and their everyday geographies; consequently, these embodied experiences of VR provided possibilities to potentially build an attachment to place in a virtual environment (Evans and Rzeszewski, 2020). The concept of place attachment is used in environmental psychology and geography to describe the emotional bonding that people develop towards a place that is meaningful for them. It is one of the components of the sense of place (other being place meaning) and it is an important element in the context of VR and live performance (Kudryavtsev, Stedman and Krasny, 2012).

However, the emergence of virtual environments that can potentially provide a space for human activity that is meaningful and has significant importance across the temporal, spatial and emotional dimensions of experience with the use of the VR technology, poses another conceptual challenge: it removes the physicality of the location from the sense of place equation. That exact physicality has been traditionally assumed to play an important role in forming an attachment to place (Lewicka, 2011). And especially in the context of live performance, physicality has an immense importance. For instance, take one of the authors of this chapter. In 1996, while attending a gig by the seminal Brazilian thrash metal band *Sepultura,* one of the authors dislocated his shoulder while in a particularly intense and physical mosh-pit. That injury did not stop the author from throwing both himself and others around with abandon, although there were some painful aftereffects. While not indicative of all live music performances, many are characterised by a physicality of proximity to others, both to the crowd and the performer(s) themselves, with that physical dimension expressed in different ways according to the performance, atmosphere, mood, and affordance of the environment. It is in that physicality that the move to live performance in VR offers fascinating divergence from the live experience in "meatspace." The close physical proximal experience of some live performance cannot be recreated in VR, but the use of the avatar in VR means there are possibilities for new spatial, emotional, and social relations to be built in the live experience that are impossible in "meatspace." Why stand with others when one can fly over the performance, observe from the stage or hover in front of the performer? The quasi-physicality of VR and the possibility of new kinds of subjectivity in live VR performance should not be instantly characterised as less than live, but instead should be seen as a new kind of being-live.

Moreover, in addition to the physicality and proximity to others of the live performance, another important element in live performance is the perception of authenticity of the live performance compared to a recording. According to Danielsen and Helseth (2016, p.24) while a live concert no longer necessarily presents a temporal or physical correspondence between the sound and vision experienced by the audience and the actual production of the event, there is a continuing promise of an experience or event that is characterised by an intimacy that affords a level of authenticity. An audience may accept a major discrepancy between the auditory and visual information provided in the performance as long as the core auditory elements are presented and received convincingly, as long as the source of authenticity is augmented by modes of authentication (Bratus, 2016, p.41). Importantly, these modes of authentication can be conveyed through an authenticity of sound and performance rather than visuality. This bodes well for the use of VR as a platform for live performance in terms of authenticity if, and only if, the aspect of intimacy can be retained for the audience. Where VR has, potentially, a distinct advantage over "meatspace" concerts is in being able to recreate this sense of intimacy beyond the recording of an event. If an embodied avatar can be sufficiently positioned in the virtual performance space, then authenticity can be achieved as a subjective mood experienced by the audience member.

Live performance in VR faces a number of challenges, but these are not dissimilar to the challenges faced to live performance in "meatspace." To achieve what we call presence, we contend that the live performance in both VR and "meatspace" must engage the audience member at three levels: the focal, the embodied, and the hermeneutic. Each of these levels is necessary but not sufficient to achieve a feeling of lived-in presence in the live performance, and we contend that while live performance in VR might be different to live performance in "meatspace," it can still achieve the feeling of presence and authenticity in the live performance by achieving engagement at these levels.

The relations of being-live in VR

The framework we develop here to understand how an authentic sense of presence in live performance in VR may be achieved is derived from post-phenomenological theory. Classical phenomenology studies our experience of the world; post-phenomenology studies how our experience is mediated by technology (Wellner, 2018). Eschewing Heidegger's concerns with Dasein, this approach retains the rejection of the Cartesian subject by replacing it with the existential, lived body as a fundamental concern (Ihde, 1993). Post-phenomenology specifically updates phenomenology by understanding the world through the *I – technology – world* schema, and thanks to this structuring of technology as a mediator of experience, post-phenomenology is also known as mediation theory. Ash et al. (2018) argue that an advantage of a post-phenomenological approach is that it interrogates how digital interfaces appear as objects, and therefore allow researchers to think about how interfaces are structured to modulate actions without reducing the modulation effect. Consequently, the approach provides an account of human experience by expanding what is meant by "human" and by re-evaluating the role of non-human objects in the construction of experience. Interfaces in this approach refer to how multiple objects work and communicate together to construct experience.

Critical to the use of post-phenomenology in understanding the differences between immersion, embodiment, and presence in VR is that embodiment replaces subjectivity in a post-phenomenological analysis, in effect creating a non-subjective phenomenology (Ihde, 2010). Action, experience, and knowledge is always situated as an embodied experience, and because of this embodiment, self-knowledge is reflexive as a factor in being-in-the-world and activity with other things as an agent in the world. In the context of VR, the kind of engaged attitude or mood that the user has through their embodied engagement with the VR space shapes their mode of engagement as immersed, embodied, or present (or none of these). The importance of the body and embodiment in post-phenomenological theory can be thought of as the human body being caught in the fabric of the world that enmeshes us, and digital technology is part of that fabric (Irwin, 2018). Stacey O'Neal Irwin (2018, p.185) terms this entanglement as the "digital attitude:" when digital technology is familiar, the body is often engaged in digital experience in a taken-for-granted manner (such as occurs in VR with body and avatar acceptance). In our average everydayness, we incorporate digital media into our own body and life experience in a habitual manner. For the understanding of VR, this approach sets a groundwork for a kind of analysis: VR is seen as part of a wider context in which our bodies and lives are enmeshed with the digital; the body is critical in all our engagements with the digital because we are embodied agents in the world, and the digital and our embodied activity act as a co-creator of our world (even if that world is virtual).

Embodiment is consequently a critical part of the post-phenomenological framework, and as such lends itself to an analysis of VR that has embodiment as a core component. The use of and digital media creates a technologically-mediated pluraculture (as opposed to having a deterministic effect), and embodiment is an element of that pluraculture as a particular form of the *I – technology – world schema* (Ihde, 2009). Recall that the schema positions the device as a co-constituent of understanding the world along with the user of the technology and the world itself. The nature of this co-construction is not uniform, however. Ihde argues that there are four relational shifts that affect the way that technology shapes relationships between humans and their world: focal relations, embodiment relations, hermeneutic relations, and alterity relations. In this analysis, we will consider how each of these relations in turn and how each contributes to a sense of presence in a live VR event. This analysis forsakes alterity relations because they refer to the quasi-otherness of relations to particular forms of digital technology, which would be incongruous with live performance and VR.

Focal relations

Focal relations refer to the phenomenon of technology in the foreground being focussed upon, while other technology works in the background. This relation is akin to a focus on a particular object, experience, or technology while that technology works in the background, away from circumspection on the part of the user. In this analysis of live VR, the focal relation is posited as immersion where the attention of the audience member is on the live performance or experience, not the technology or the inter-connectivity of the technology with other technical devices or the functioning of hardware or software in the background. Attention in a focal relation can be multisensory rather than just visual, although in VR the visual may be the primary mode of focus. This focus on attention is critical to a sense of presence in the live performance in the same manner that such a focus would not necessarily be needed in a live performance in a "meatspace" concert.

The notion of a focal relation corresponds with Bortulussi and Dixon's (2003) definition of immersion as a hybrid, dynamic and interactive phenomenon that involves convergence and divergence to the state of immersion. Such a view emphasises the role of the individual in the construction of immersion, as immersion involves an orientation towards engagement with the media in question. Thon (2008) positions this as a kind of attentional focus, a psychological immersive shift of attention that goes hand-in-hand with the construction of situational models of engagement. Ryan (2015) furthers this notion of the psychological aspect of immersion by arguing that immersion is a kind of directed, intentional consciousness that relates to another world and reorganises the "universe of being" around that world.

Embodiment relations

The embodiment relation is more significant to this discussion and expressed as:

(I – technology) world

Arrows and parentheses allow for permutations on the I – technology – world schema, and arrows denote intentionality towards that unit of the schema (Wellner, 2018). Selinger (2006) describes embodiment relations as the relation that occurs when we use technology to amplify the body's perceptual abilities (in a sense, a McLuhanist extension of man). In an embodiment relation, the technology is always in a ready-to-hand state, being used to perceive the world through the technology itself. We have embodiment relations with many technologies, from eyeglasses to television. Ihde (2010) argues that computers have an embodiment relation to users because we use computers in our everyday understanding of the world; computers are part of the fabric of the everyday and our interactions with them are shaped by an intentionality towards the world. Embodiment is, therefore, always limited by the scope of the program being used and our embodiment relations are always inter-relational (Ihde, 2010). The embodiment relation in our analysis of live performance in VR translates as embodiment directly. When using VR, the technology itself is embodied (typically through a device worn on the head of the user), and the world is being "seen" or experienced through that technology, with bodily motion and perceptual attention being mapped by the VR equipment and fed back to the user as an experience that has visual fidelity and congruence. Additionally, we embody an avatar that affords us a level of physicality in the virtual environment that is an element of the embodiment relation and crucial to presence. It is here that there is the clearest (for now) difference between living in VR and living in "meatspace."

Hermeneutic relations

More significant again is the hermeneutic relation. In a hermeneutic relation, we are perceiving the world through the technology itself.

I (technology – world)

The interpreting or reading of the world is through the technology. Hermeneutic relations involve entering into practices with artefacts to gain knowledge of the world otherwise not available (Selinger, 2006). Wellner (2018) argues that hermeneutic relations are our main post-phenomenological vehicle to understanding our special relation to media. Feenberg (2006) describes hermeneutic relations like "a screenplay in which the interpreted message is, in effect, a world." Feenberg's point is that in a hermeneutic relation, technology and the world are not just two parts of a unit – technology-world as a co-entity replaces the world as a focus of our intentionality. This relation is mapped to the feeling of presence in VR. In a hermeneutic relation, the world and technology form a singular unit where intentionality is directed. The hermeneutic relation involves a translational mediation of technology and technological codes back to the human, but those codes and the technology are a co-constructor of the world in which the user is an active participant. Therefore, in a hermeneutic relation we feel "present" in the experiential world that has been co-created by the technology.

The notion of technology as a co-constructor of world leaves some room for a technologically determinist critique of the theory, but the importance of intentionality on the part of the human and the technology itself as a co-constructors is a means to avoid this critique. In this view, "intentionality is not a bridge between subject and object, but a fountain from which the two of them emerge" (Rosenberger and Verbeek, 2015, p.15). Consequently, technological intentionality is a co-constructor of the world rather than a determinist shaper of the world, as the intentionality on the part of the user towards the (technology – world) unit is as critical as any intentionality programmed into the technology itself. Wellner (2018, p.215) proposes a new relation to explain this understanding:

I (technology world)

The intentionality of the technological artefact to represent the world in a manner where the world itself becomes a part of the (technology – world) unit does not detract from there needing to be a human being with intentionality directed to that unit in order for the hermeneutic relation to exist and for there to be a hermeneutic effect. In other words, the world is read through the technology and is therefore contingent upon the technology to be meaningful as a world. Intentionality plays a critical role in the establishment of presence in VR, as it does with the feeling of presence in live performance. The intentional nature of how these performances should be experienced and embodied defines the live experience and possibility of presence as an elevated and deep engagement with the VR live performance – if you don't want to feel it, then you will not feel it.

Presence from relations

This analysis has mapped the three relations onto the three states of engagement with VR:

Immersion	-	Focal relation
Physicality or embodiment	-	Embodiment relation
Intentionality in world or environment	-	Hermeneutic relation

The contention of this analysis is that these relations should not be read as discreet; instead, as O'Neal (2018) argues the schema can be laid on one another at the same time. Presence in VR (and at a live performance in "meatspace") is therefore a combination of the focal + embodiment+ hermeneutics relations, where the hermeneutic relation is critical but cannot be achieved without focal immersion and embodiment. The framework also allows for shifts between these different states of engagement with the VR experience. Ihde (2009, p.190) argues that relations between humans and technology shift thanks to five potential variational distinctions: materiality of the technology, bodily technique of use, cultural contents of the practice, embodiment in trained practice, and the appearance of differently structured lifeworlds. The variations between different relations occur due to pivot points. A pivot "stresses the degree to which the material of the artefact and human attentions can create different uses" (Whyte, 2015, p.76), with the movement between pivots and stabilities creating multistabilities of use. The structure of technologies is multistable with regards to use, cultural embeddedness, politics and ethics, and is critical to the sense of place and placehood that emerges from the use of technology.

Conclusions: Live-ness in VR and "meatspace"

Our analysis leaves us with a comparison between being-live in "meatspace" and being live in VR:

Table 8.1: Types of relationship in "meatspace" and VR live performances.

Relation	"Meatspace"	VR
Focal	Attention to the act	Attention to the act and the technology
Embodiment	Physicality, proximity to performance and others	Embodiment of avatar, proximity to performance in VR space
Hermeneutic	Intentional relation to the act	Intentionality, world as meaningful reality
Combination	Presence1	Presence2

Authentic presence is a function across these relationalities, but our analysis indicates that the kind of presence that can be achieved in VR will be different to that in "meatspace." VR experience, with its unique affordances, can be an important part of the focal relation; however, the hapticity and physical embodiment of VR avatars is considerably different from being physically embodied in "meatspace." Consequently, it is this dimension that has the most striking difference between the two arenas for live performance. That is not to say that an audience member cannot be embodied in VR, but that embodiment is qualitatively and experientially different to that in "meatspace."

Following the notion of the pivot described above, the pivot that differentiates the VR live experience and the "meatspace" live experience is the embodiment aspect. It is this pivotal aspect of experience which leads to a kind of presence (presence[2] above) that is demonstrably and experientially different to the live performance in "meatspace" – not necessarily worse, but different. Both require a focal aspect, an embodied aspect and a hermeneutic or intentional aspect to achieve a sense of presence, but the pivot means that for now these senses of presence will differ considerably. In this context, live performance in VR does not replace live performance because it is a different form of performance itself that can lead to presence (although may not) at the performance, but the presence will remain different from live performance in the physical world. Confusing the two is not helpful; acknowledging the difference and taking advantage of the unique affordances of VR rather than trying to directly replicate physical performances will likely be a more successful path towards building engaging, embodied "liveness" in future forms of VR.

References

Ash, J., Anderson, B., Gordon, R. and Langley, P. (2018). Unit, Vibration, Tone: A Post-Phenomenological Method for Researching Digital Interfaces. *Cultural Geographies* 25 (1): pp. 165–81.

Atarashi, R., Sone, T., Komohara, Y., Tsukada, M., Kasuya, T., Okumura, H., Ikeda, M. and Esaki, H. (2018). The software defined media ontology for music events. In *Proceedings of the 1st International Workshop on Semantic Applications for Audio and Music* (pp. 15-23).

Augé, M. (1995). *Non-places: Introduction to an Anthropology of Supermodernity.* Verso.

Berkowitz, Z.A. (2016). Music in Virtual Space: Theories and Techniques for Sound Spatialisation and VR-Based Stage Performance.

Bratus, A. (2016). In-between Performance and Mediatisation: Authentication and (Re)-Live(d) Concert Experience, *Rock Music Studies*, 3:1, pp. 41-61.

Brown, S.C. and Knox, D. (2017) 'Why go to pop concerts? The motivations behind live music attendance', *Musicae Scientiae*, 21 (3). https://doi.org/10.1177/1029864916650719

Charron, J.P. (2017). Music Audiences 3.0: Concert-Goers' Psychological Motivations at the Dawn of VR. *Front. Psychol.* 8. https://doi.org/10.3389/fpsyg.2017.00800

Chidekel, D. (2021). The Future of Live Events: AR, VR and Advertising. *Forbes*, 09/02/2021. https://www.forbes.com/sites/legalentertainment/2021/02/09/the-future-of-live-events-ar-vr-and-advertising/ (Accessed September 1 2021).

Couldry, N., and Hepp, A. (2018). The continuing lure of the mediated centre in times of deep mediatisation: Media Events and its enduring legacy. *Media, Culture & Society* 40, pp. 114–117. https://doi.org/10.1177/0163443717726009

Danielsen, A. and Helseth, I. (2016). Mediated Immediacy: The Relationship between Auditory and Visual Dimensions of Live Performance in Contemporary Technology-Based Popular Music, *Rock Music Studies*, 3(1), pp. 24-40.

Evans, L. (2015). *Locative Social Media: Place in the Digital Age*. Palgrave.

Evans, L. (2018). *The Re-Emergence of VR*. Routledge.

Evans, L. and Rzeszewski, M. (2020). Hermeneutic Relations in VR: Immersion, Embodiment, Presence and HCI in VR Gaming, in: Fang, X. (Ed.), HCI in Games, Lecture Notes in Computer Science. Springer International Publishing, Cham, pp. 23–38. https://doi.org/10.1007/978-3-030-50164-8_2

Feenberg, A. (2006). Active and passive bodies: Don Ihde's Phenomenology of the Body in Selinger, E. (ed.) *Postphenomenology: A critical companion to Ihde* pp. 194. SUNY Press, Albany.

Gao, Z., Lee, J.E., McDonough, D.J., and Albers, C. (2020). VR Exercise as a Coping Strategy for Health and Wellness Promotion in Older Adults during the COVID-19 Pandemic. *Journal of Clinical Medicine* 9, p. 1986. https://doi.org/10.3390/jcm9061986

Geigel, J. (2017). Creating a theatrical experience on a virtual stage, in: International Conference on Advances in Computer Entertainment. Springer, pp. 713–725.

Halegoua, G.R. (2020). *The Digital City*. NYU Press, New York.

Hammelburg, E. (2020). Live event-spaces : Place and space in the mediatized experience of events, in: *Locating Imagination in Popular Culture*. Routledge, pp. 215–229. https://doi.org/10.4324/9781003045359-17

Heidegger, M. (1962). *Being and Time*. New York: Harper and Row.

Horie, R., Wada, M., Watanabe, E. (2017). Participation in a VR concert via brainwave and heartbeat, in: International Conference on Applied Human Factors and Ergonomics. Springer, pp. 276–284.

Ihde, D. (2009). *Postphenomenology and Technoscience: The Peking University Lectures*. SUNY Press, Albany.

Ihde, D. (2010). *Embodied technics.*_University of New York Press, NYC.

Irwin S.O. (2018). The Unbearable Lightness (and Heaviness) of Being Digital. In: Romele A., Terrone E. (eds) *Towards a Philosophy of Digital Media*. Palgrave Macmillan, New York.

Janer, J., Gómez, E., Martorell, A., Miron, M., de Wit, B. (2016). Immersive orchestras: audio processing for orchestral music VR content, in: 2016 8th International Conference on Games and Virtual Worlds for Serious Applications (VS-GAMES). IEEE, pp. 1–2.

Kharouf, H., Biscaia, R., Garcia-Perez, A., Hickman, E. (2020). Understanding online event experience: The importance of communication, engagement and interaction. *Journal of Business Research* 121, 735–746. https://doi.org/10.1016/j.jbusres.2019.12.037

Kudryavtsev, A., Stedman, R.C. and Krasny, M.E. (2012). Sense of place in environmental education. *Environmental education research*, 18(2), 229-250.

Lewicka, M. (2011). Place attachment: How far have we come in the last 40 years? *Journal of Environmental Psychology* 31, 207–230. https://doi.org/10.1016/j.jenvp.2010.10.001

Manavis, S. (2021). How the Pandemic Made VR Mainstream. New Statesman, 02/25/2021 https://www.newstatesman.com/science-tech/coronavirus/2021/02/how-pandemic-made-virtual-reality-mainstream retrieved 09/07/2021.

Muhanna, M.A. (2015). VR and the CAVE: Taxonomy, interaction challenges and research directions. Journal of King Saud University-Computer and Information Sciences 27, 344–361.

Relph, E. (1976). *Place and Placelessness.* London: Pion.

Rosenberger, R. and Verbeek P.P. (2015) *Postphenomenological investigations: essays on human-technology relations.* Lexington Books, Lexington.

Ryan, M. (2015). *Narrative as VR 2: Revisiting immersion and interactivity in literature and electronic media.* John Hopkins University Press, Baltimore.

Selinger, E. (2006). Postphenomenology: A critical companion to Ihde. SUNY Press, Albany.

Slater, M. (2003). A note on presence terminology. PRESENCE—Connect. URL http://presence.cs.ucl.ac.uk/presenceconnect/articles/Jan2003/melslaterJan27200391557/melslaterJan27200391557.html (Accessed: 8 February 2020).

Slater, M., Pérez Marcos, D., Ehrsson, H., Sanchez-Vives, M.V. (2008). Towards a digital body: the virtual arm illusion. Front. Hum. *Neuroscience.* 2. https:// doi.org/10.3389/neuro.09.006.2008

Tarumi, H., Nakai, T., Miyazaki, K., Yamashita, D., and Takasaki, Y. (2017). What do remote music performances lack?. In *International Conference on Collaboration Technologies* (pp. 14-21). Springer, Cham.

Thon, J. (2008). Immersion Revisited. On the Value of a Contested Concept. In O. Leino, H. Wirman, and A. Fernandez (Eds.), *Extending Experiences: Structure, Analysis and Design of Computer Game Player Experience* pp. 29-43. Lapland University Press, Rovaniemi.

Tuan, Y. F., 1975. Place: An experiential perspective. *Geographical Review*, 65, pp. 151-165.

Wellner G. (2018). From Cellphones to Machine Learning. A Shift in the Role of the User in Algorithmic Writing. In: Romele A., Terrone E. (eds) *Towards a Philosophy of Digital Media* pp. 215. Palgrave Macmillan, New York.

Whyte, K. (2015). What is Multistability? A Theory of the Keystone Concept of Postphenomenological Research. In Technoscience and Postphenomenology: The Manhattan Papers, ed. Jan Kyrre Berg O. Friis and Robert Crease, pp. 76. Lexington Books: Lanham.

Yakura, H., Goto, M. (2020). Enhancing Participation Experience in VR Live Concerts by Improving Motions of Virtual Audience Avatars, in: 2020 IEEE International Symposium on Mixed and Augmented Reality (ISMAR). Presented at the 2020 IEEE International Symposium on Mixed and Augmented Reality (ISMAR), IEEE, Porto de Galinhas, Brazil, pp. 555–565. https://doi.org/10.1109/ISMAR50242.2020.00083

Chapter 9

The Under Presents: The potential of the real in virtual theatre

Alicia Corts

Saint Leo University

Abstract

VR theatrical performance is in its nascent stage, yet the possibility for expanded, immersive performance exists within these spaces. Using Lawrence Halprin's RSVP concept for devised theatre, this chapter looks at the ways in which three virtual performances—*Carney y Arena*, *Collider*, and *The Under Presents: The Tempest*—demonstrate how this expansion of performative resources must include consideration of the virtual and physical bodies of both audience and performers. By reaching beyond the physical world limitations of biology, immersive theatre can now tell stories in ways that challenge the audiences' perception of the world and their place in it.

Keywords: virtual performance, phenomenology, virtual theatre, immersive theatre

***

Introduction

The combination of VR and theatrical performance is a marriage that is far from new. Dating back to the late 1990s, academics and artists experimented with the merging of technology and artistry to explore the possibility of audience immersion in VR. Here, early attempts focused on a seamless delivery of video, an experiment that required the use of supercomputers to merge images (Boswell and Gardner, 2001). Likewise, theatre companies have used YouTube as a repository for performance documentation as 360-degree cameras have become more accessible. These videos give the user a sense of "being there,"—witness to a performance that has already occurred. Active engagement as it pertains to the audience, however, ends here. The audience does not influence the performance in any way; they are there merely to observe.

This chapter deals with a different kind of theatrical experience in VR: immersive theatre. With the rise of VR headsets, especially the standalone Oculus Quest, which does not require a powerful standalone computer to function, theatre companies have begun experimenting with productions that mimic the kind of immersive theatre of Punch drunk's *Sleep No More*, a project where audience participants roam through a space where they can interact with characters and experiences. Rather than merely imitating these physical world performances, however, virtual immersive theatre productions can challenge our notion of what both the physical and virtual body can experience. These performances, which pay close attention to scenography, build worlds that challenge the audience to explore rather than just to witness. As Marvin Carlson (2013) describes it, VR performances have a sense of "here-ness," a distinct sense of being present in a space with other audience members and performers. Drawing on Lawrence Halprin's (1970), and Carol Martin's (2012) notion of realness in the context of phenomenology, I will examine why the immersive theatre of VR presents a new way of interrogating and utilizing virtual bodies.

To begin to understand how bodies become a unique theatrical tool in VR, I open with a brief history of both immersive worlds as well as theatre's form of immersion, followed by a mapping of how the body is asked to operate in our current VR spaces. Finally, I examine three productions—*Carne y Arena*, *Collider*, and *The Under Presents: The Tempest*—in order to draw conclusions about the place of both the physical and virtual bodies.

Immersive spaces

As far back as the eighteenth century, artists have produced spaces where audiences could feel immersed in a particular location at a particular point in time. Robert Baker, a Scottish painter, premiered the first panorama in Edinburgh in 1788 to great acclaim (Piccolin, 2007), so much so that the panorama became the new, fashionable entertainment in London through the nineteenth century (Boseman and Kang, 2001). These panoramas were massive in size, and depending on the viewing space, often immersive: their main purpose was to change a visitor "from a passerby to an eyewitness of highly significant events" (Sternberger and Neugroschel, 1977, p. 4). These panoramas featured no actors to bring the scenography to life, yet they were performative in the way they drew in the audience to *feel* part of an event. The painters and operators designed these rotundas so that the audience began to experience a sense of presence in the scene itself: in Paris, for example, the "Le Vengeur" panorama rotunda—an exhibit that featured the naval battle of Queffant—featured a hydraulic floor and mimicked the movement of the sea as the audience observed the painting, giving them a sense of being on the bow of one of the other ships (Piccolin, 2007). Robert Barker, in fact, made it clear in his patent for the panorama that the purpose of the exhibit was to "make observers, on whatever situation he may wish they should imagine themselves, feel as if really on the very spot" (Ellis, 2008, p. 134).

Carlson (2013) stresses that theatre has two great components that must exist for the audience: a sense of the event happening "now" and "here." Those two components, as noted by Robert Barker in his patent, is something that the audience imagines for themselves. If they feel they are present in the action, the historical moment can feel like "now" and "here." For the audience members standing on the moveable platform for the "Le Vengeur," the moving floor and the dramatic painting gave them the tools they needed to believe they were at the Battle of Queffant. There is power in the ability to move an audience by making them feel as though they are present for the event. Carlson (2013, p. 19) notes that the power comes from the "separation of the real world of the spectator from the mimetic world of the actor" and the delicate balance the audience achieves in navigating between those spaces. The person standing on the platform for "Le Vengeur" knows they have just come from the streets of Paris, and they are also aware that they have entered a space of performance. They willingly choose to believe they are in the midst of a battle that they most likely did not experience in its original moment in time. The panorama draws them into a space where they can suspend disbelief because the environment leads them to such a conclusion.

The Walt Disney Corporation took advantage of the popularity of panoramas as the era of film launched in the twentieth century. Ub Iwerks, one of the masterminds behind the Disney empire, created the Circle-Vision 360 camera, an apparatus housing nine cameras in a circle intended to capture a true 360 experience (Gennawey, 2014). In a 1984 internal publication, the Disney Corporation declared that Circle-Vision 360 doesn't just tell a story "but also communicates a sense of involvement, learning and a whole range of human feelings" (Walt Disney Corporation 1984, p. 31). As the medium for these panoramas moved from still painting to moving images on film, the investment for the audience increased. Audiences no longer had to imagine that people were moving in the images around them but could see and experience that movement for themselves. Tony Bennett (1995, p. 241) notes that the shift toward film made these events more like the thrill rides in the rest of the Disney Parks: "whereas thrill-rides take the normally stationary body and hurtle it through space, they hurtle the vision through space while fixing the body as stationary." There is a sense of movement, which is why in Disney auditoriums for these films there are safety rails that people can lean on if they become too overcome with the motion of the film. For Disney, giving people this cinematic ride through a different environment was a means to propagate a distinct vision of America during the Cold War: the company recognized that the audience was more invested in these types of presentations, so they used the

Circle-Vision 360 presentations as a way of showing the world the "supposed rightness" of the American lifestyle (Nilson, 2011, p. 250).

VR headsets follow in the path of these technologies by immersing audiences in a created space. The connection is easy to see in apps like *Wander* for the Oculus Quest. The Quest uses panoramic pictures taken for Google's Street View feature and places the user in the centre of the panorama. The user can simulate moving through the map by clicking on the next panorama, but they cannot interact or touch objects within the photo. Like the audiences of old staring at a panorama, these users can turn in their stationary position, looking at images that surround them. Similarly, 360-degree video in spaces like YouTube invite the VR user into a more intimate version of the Circle-Vision 360. These environments certainly fulfil the sense of a user being present in a place and a particular moment in time—what Carlson says is necessary for theatre performance. Yet, these attractions do not have the intimacy of the theatre. So, what is missing?

I propose that the key to the missing performance is the full, physical body of the user being recognized in the space itself. In *Wander*, for example, there is no representation of the user's physical body in the digital space of the app. Similarly, in 360 videos, the camera is representative of the user's body rather than there being a solid object to see. When the body is represented in VR games like *Red Matter* or exercise apps like *Holofit*, the body is not fully realized. *Red Matter* only displays the user's hands, while in *Holofit*, the whole body is represented but isn't in sync with what the physical world body is achieving. Let me use an example to illustrate this point. When riding a bike, for example, the virtual body may be cycling slower or faster than the user knows their physical body is achieving. This disconnect between the physical and virtual body doesn't indicate that meaning can't be made without the presence of a virtual body. Susan Kozel (2007) looks at these types of entertainments as a means for an audience to make sense of their world and environment. She argues that the physical body itself experiences the shift in cultural, social, and existential paradigms and requires a loop of self-reflection to sort out how that body—or in this case, the reflection of the physical body—will fit into the changing world. Panoramas brought to the public a closed performance of self-reflection that involved substantiated, phenomenological presence within a performance space, what Kozel (2007, p. 11) would call the "engagement between bodies and machines within the world" to make meaning. When a body interacts with a mechanism—like a panorama, a Circle-Vision 360 film, or a VR apps like *Wander*, *Red Matter*, and *Holofit*—the user can project their own sense of self into that moment as a means of self-reflection. Their mind is active in this process, but their body is not working toward or only partially mimicking the reflection. The user is only somewhat in the here and now, with the constant reminder that their avatar body—whether invisible in *Wander*, simply hands in

Red Matter, or an out-of-sync reflection in *Holofit*—is in a different 'here' than their physical body.

Where we do see the merging of environment, time, performance, and body is in the world of immersive theatre. Gareth White (2012, p. 222) defines this type of performance as "ready-made exploratory landscapes, redolent of other histories, into which performances can be scattered, and in which engagement with the environment can be an important part of the audience experience." The engagement with the environment is what makes these performances unique. Punchdrunk's *Sleep No More*, one of the more famous of these productions, took over three warehouses in New York to create a multilevel performance environment where individual rooms hosted snippets of performance all based in the world of Shakespeare's *Macbeth*. Here, audience members made choices about where to wander, what food and drink to consume, which performers they wanted to interact with, and how they wanted to respond. In doing so, the audience bodies became a vital part of the performance itself, and more importantly, the performers saw and acknowledged those audience bodies within that space.

The body in VR

Since the audience and their virtual bodies become important in VR performance, they become important resources to consider when planning performances. In Lawrence Halprin's (2014) conception of the RSVP method of devising, we can begin to see why the body is a necessary component for immersion in VR. RSVP stands for resources, score, valueaction, and performance, and the four components influence each other to create the final outcome. Consider *Sleep No More* in light of these four properties. In terms of resources, the show not only had an immense amount of space but lighting equipment, performers, sets, costumes, and audience members as resources. The score refers to the way that the audience moves through a space, and in *Sleep No More*, the audience was at times guided (closed score) and were also allowed to wander freely (open score). The mix of the two scores helped guide audience members into the story while also keeping them a part of the production. Valueaction is how you want the audience to respond based on a particular value. Felix Barrett, one of the two directors of *Sleep No More*, stated that he "wanted to create work that would have a physical impact on [an audience member's] body" (Bakchormeeboy, 2019). When constructing the performance, therefore, the resources are arranged in the score so that the audience is moved to act in a specific way. In *Sleep No More*, each room contained items that audience members might examine, such as letters, diaries, and even pieces of candy to eat, using their bodies to explore the space. Finally, the performance puts all the elements into action, which gives the creators the ability to see the work in action and adjust the other areas if necessary.

One of the barriers to immersive performance in the physical world is the sheer magnitude and cost of such an adventure: a resource issue in RSVP terms. The budget for the Shanghai version of *Sleep No More* had a budget of twelve million dollars, far beyond the capabilities of most theatre companies and artists (Shuangjie, 2016). Another barrier is the safety of the audience as they move through space: most audience members wouldn't consider participating in a show where one person would be dropped off a cliff, but VR allows for the freedom to let an audience feel that kind of danger by placing their virtual bodies in just such a set of circumstances. Audience bodies are a resource, but they bring with them a degree of responsibility for their wellbeing. Finally, immersive theatre is also specifically located in one place. Audiences can go to the *Sleep No More* site to participate, but they cannot magically appear in front of the pyramids of Egypt. The resources of the physical world bring with them challenges that make immersive experiences seem daunting because of their limitations, but the virtual world opens the number of resources available for theatre artists.

The body, however, is also the starting place for experience, and immersive theatre creates worlds where experience can happen. For Alston, these situations constitute "experience machines," which he defines as follows:

> Experience machines are enclosed and other-worldly spaces in which all the various cogs and pulleys of performance—scenography, choreography, dramaturgy, and so on—coalesce around a central aim: to place audience members in a thematically cohesive environment that resources their sensuous, imaginative and explorative capabilities as productive and involving aspects of a theatre aesthetic (2016, p. 2)

The cohesion is where the importance of merging resources, score, and value action ahead of performance lies, and VR gives us the tools to open those resources for creators. Those productions might use VR as an extension of the physical world reality or as the creation of a world where fantastic realms can exist (Seamon, 2018). In any of those cases, the goal is to bring, in Alston's terms, a sensuous performance, one that utilizes the physical and performative bodies.

The key to the experience is the way that body of the user moves through the space, and VR gives an overabundance of resources to structure any space imaginable. Because VR takes away restrictions on place and time, the user is untethered from the reality of the physical world in any performance in the virtual space (Kelly, 2017). If a performance starts in a theatre, the audience and performers can just as easily be transported to a field in Chile or a castle in Germany. This shift represents an explosion in the number of resources available to the theatre. No longer is the design or conception of the show based solely on monetary resources. Borgmann (1992, pp. 87-102) suggests that the physical world "encumbers and confines" while the virtual space is "pliable" and "brilliant," and infinitely more capable of experiences we have yet to dream of.

The physical experience is the key to immersive theatre and VR. Alston's experience machines ask the audience for their productive participation: rather than sitting by and observing, they must interact with the performers and the environment for the performance to enhance their own sense of being a part of the world. Just as this productive participation is important for immersive theatre in the physical world, theatre in the virtual space must consider how the audience participates in each production. Crafting virtual theatre with productive participation in mind asks the user to use their physical and virtual body within the virtual space to actively use their perception of the environment and get feedback for their efforts. Merleau-Ponty (2013) in *Phenomenology of Perception* emphasized that the body understands the world through its awareness of its surroundings, since the body is the epicentre of intention flowing back and forth from the user and the world. Productive performance ties our virtual and physical bodies together in a phenomenological self-reflective loop of experience. When my virtual hand reaches out to touch a performer or player who responds to that touch, my brain perceives the experience as reality, even if the space is virtual. This perception collapses the distinction between the physical and virtual worlds, allowing the user to experience even the most outrageous of worlds as real:

> Though VR may superficially seem real, it can readily escape from and replace the lived messiness of real life worlds with more convenient, vivid, or fantastical situations that require no stakes, answerability, or efforts of will.... Human beings are always already immersed in their worlds, even if that immersion becomes virtual. (Seamon, 2018, p. 12).

Returning to the concept of RSVP, the resource of an audience member's experience is vital to any virtual theatrical performance. Moving through a virtual world with no other players or viewing a panorama in an app like *Wander* asks the user for non-productive action: the user may turn to see the entire panorama or use the controls to move through a space, but nothing about the world or the storyline changes. Within the phenomenology of theatre, experience springs from where the subject is positioned and how they are allowed to interact with the story and their surroundings (Raynor, 1994). The difference lies in the level of immersion: does the audience member's body change the environment it finds itself in?

Martin (2013) presents the notion of the injection of the real into theatrical performances, and once again, the audience member's subject position is important and critical. Imagine the lights go out in a play, accompanied by sounds that indicate an electrical failure of some kind. Is the outage part of the play? The audience member is suddenly thrust out of a complacent mode into a more active one: they must decide if they will seek help, wait in their seat, or even scream in terror. The real can also be a bridge to reality. Martin (2013) explains that the audience recognizes acting as a replacement for a fully-lived, real life, yet when a real person is playing themselves onstage—such as Spalding Grey's autobiographical monologues—and when clear evidence is offered that the person onstage is, in fact, real, the distinction between theatre and reality blurs.

In the virtual space, most apps and games have interactions with NPCs, or non-playing characters, which often give clues to the game or move the player forward in the story. These NPCs have a finite array of responses, so while a player can take time to interact with them, the experience becomes repetitive after the player has exhausted the number of interactions. The player's experience with these characters may lead them to other actions in the game, but the player cannot influence the NPCs or draw them away from their standard responses. The player can only self-reflect on their own actions without considering the performance of the NPC. In some games, a player may be required to shoot an NPC, yet the reflection might be more of how that action advanced them in the game rather than on the violence of the action. After all, the NPC is not an embodied character: there is no evidence that they are real in their performance.

What VR theatre can potentially accomplish is to draw the user's body back into the experience by injecting the reality of an embodied performer or other audience members interacting with user in real time. When someone, for example, compliments my avatar on their ability to move through space effectively, I know that the avatar/user complimenting me sees the movement of my physical body to control my virtual avatar in real time, making my performance something that embodies Carlson's "here" and "now" in a way that is recognized by others. When using an app like *Wander*, while there is a potential for self-reflection, there is no feedback that suggests others can see my choices. If I click to move in Wander, no one challenges my choices. In *Red Matter*, elements of the game world—a long-abandoned science station—give me feedback on my playing, such as when I push a wrong button in a puzzle. Yet the game itself never asks me why I decided to make the choice to dance about the station rather than playing a puzzle. It is concerned only with my choices in relation to the gameplay, not in how I react to the environment and perform outside of the game's storyline. Immersive VR theatre brings the promise of interactive stories and interactions that meld the reality of both the physical and virtual worlds. In the following examples, users' physical and virtual bodies meld because of those interactions.

The body in production

Carne Y Arena

Carne y Arena, an Oscar-winning filmed installation by Alejandro G Iñárritu, demonstrates one type of immersive virtual theatre. The installation is a carefully motion-captured recreation of a refugee's journey to a new land, placing the audience participant in the scenario where American soldiers discover a group of refugees in the middle of the night. Each audience member can wander freely through the experience of being stopped by the American soldiers. Subtitled "Virtually present: Physically invisible," this production brings three audience members each into separate rooms to experience the performance simultaneously. Each barefoot participant wears a headset and earphones, and the large performance spaces have floors covered in sand.

In this production, then, the audience member is put into a specific physical-world location where the scenography of the space gives them the bodily feedback to become fully immersed. The visceral feeling of sand creates the feedback loop necessary for the audience member to connect their physical sensation with what they see in the virtual space. The performance begins with audience members being led into a concrete room with metal benches, on which there are various objects confiscated from refugees at the United States border. The audience can interact with the objects in this space before heading into the VR room, one-by-one. Two guides carry the cables necessary to operate the audience member's headset as well as keep the audience away from the danger of the solid walls. The effect of the sand remains a constant reminder of the audience member's body, connecting them to the experience. The sand is the same in the film and in the physical room, and the viscerality provides a sense of realness into the exploration of the virtual space, reminding audience members of the concreteness of the physical room they are in as well as the virtual situation.

Carne y Arena works as an experience in the same way that a thrill ride would in a theme park like Walt Disney World. Audiences go through a preparatory space that sets the tone, they experience the ride, and then they are left to ruminate on the experience. We can see this type of construction in Disney rides: King quotes a travel writer who declares that "you don't go to the Disney parks to ride the rides. You go for a journey of the mind" (King, 1993, p. 55). Take Space Mountain, for example. In the line leading to the ride, there are games and interactive displays that propose the rider is about to go into space. The scenography of the ride reinforces that assumption, with the cars looking like spaceships that are cleared for launch. After the ride in the darkened room made to look like outer space, the spaceship returns to the launching area, and the rider exits through a hallway toward a store with space-themed products.

This Disney experience results in something that resembles a journey outside of time and space, one that favours the set, costume, sound, light, and media design more than the ride itself. In this case, the potential for the performance is in the way audience members can be present for situations outside of time or ability. The political overtones of *Carne y Asada* are obvious, but its use of resources to tell that story is unique. By fusing the physical world elements of scenography in the first room with the virtual space in the second, the separate experiences merge into a single experience for the user. The pattern for the thrill ride type of VR performance rests in how it is arranged. The VR component of the experience is last, the "ride" component. The audience is guided through a curated, scenographically-rich experience prior to putting on a headset to enter a different time, place, or situation.

Collider

To be clear, *Carne y Arena* uses an enormous amount of physical world space, money, and performer presence. *Collider,* a performance created by Single Thread Theatre Company for the Oculus Quest app *AltVR,* uses only the resources within the virtual space to create a performance. This efficient use of virtual resources is a marked difference. The performance takes a group of audience members on a tour of an abandoned supercollider that experienced a complete malfunction. The audience walks with members of the theatre company through the space, encountering child-like, alien avatars transformed by the accident. The damaged inhabitants of *Collider*— their skin glowing with a variety of colours—demonstrate for the audience how to manipulate objects, jump through the space, and where to stand when some of the less savoury characters in the supercollider world decide to fight it out. The space is expansive, with ceilings hundreds of feet in the air, walls made of lava, and virtual audience members occasionally dropping hundreds of feet to different floors to access different parts of the performance. The alien creatures move through ever-increasing areas designed to represent larger-than-life vistas: an enormous ice cave with a frozen lake as the floor, an abandoned subway line, a destroyed laboratory, and a thirty-story warehouse with atoms and pixels dancing to the hum of the ailing supercollider. The performers spent weeks in the AltVR app, building the space, devising the performance, and specifically looking for ways to consider how to transform the virtual space of performance (*Collider* website, 2021).

There was a sense that the spaces themselves drew the audience toward the next part of the story. The spaces were designed so that there was a natural inclination to move to the next room, such as slanting floors or narrowing of walls toward the next door. In other rooms, such as the ice palace, the shape of the room demonstrated the path the audience needed to follow: the ice palace was in the shape of an L, for example. Most intriguing, however, were the ways in which the space itself seemed to move both performers and audience members. At the end of the performance, all the participants were hustled through a small opening in the floor of the lava room as the villain threatened to capture audience members. There was an urgency on the part of the performers to make sure each virtual audience member made it safely into the shelter. As each person made it to the opening, they dropped into the new space, falling what would have been about fifteen feet in the physical world. The performers and audience members alike were palpably relieved when everyone made it through the opening. The environment created that moment within the performance and through the shared emotion that followed.

This acceptance and reaction to the environment by both performers and audience represents *Raumempfindung*, the Bauhaus concept of a total space where all the participants within a theatrical space react to and are acted upon by the environment as equals (Fisher, 2018). Where *Carne y Arena* made the audience's body disappear, *Collider* used the bodies of performers and audience members within the environment to not only push the story forward but to also make the story personal and important. The bodies of all of those present in the performance, therefore, are important to virtual theatrical performance, and the more the fleshly form is considered, the more immersive the performance will be. Physical scenography considers all aspects of the performance to produce a theatrical world where audience engagement and reception is key (Oddey and White, 2006). Virtual scenography has the new resource of the virtual body of the performer as well as the audience members. Virtual designers can decide, for example, to have all audience members be ten-feet tall with performers who are half that size. Biology is no longer a barrier to the resource of the human body in the virtual space.

The Under Presents: The Tempest

The performance that comes the closest to the Bauhaus ideal is *The Under Presents: The Tempest*, a true open-scored devised theatre production. Throughout the performance, audience participation is not mandated; rather, they are allowed to explore a post-apocalyptic landscape in a way that casts new light on the Shakespearean work. The audience affects the story, the environment, and the performers, just as the other elements work upon the audience to dissolve the physical body and transport them to a fictional world. This open exploration is limitless in the virtual space, something a physical-world theatre producer would find constricting. Given the expansion of resources, the "R" of the RSVP cycle, theatre suddenly has a wider toolbox to imagine not just the body of the actors but the bodies of the audience as well.

The Under Presents is an app for the Oculus Quest. The surreal world of the app leads users to a story of a forgotten shipwreck, one with secrets for the user to uncover. The user has a virtual body that they control with the hand controllers of the Quest that resembles the abstract figure of a woman on a bathroom stall door, a geometric collage giving the illusion of a body that glides through the space with the user's commands. One intriguing part of *The Under Presents* is that the game features occasional live performers wandering through the world as characters, giving feedback to the players as they progress through the game.

At the start of the novel coronavirus pandemic in 2020, the actors who worked within *The Under Presents* proposed crafting an immersive version of Shakespeare's *The Tempest.* The performance begins with each audience member finding their way to the lobby of an abandoned theatre. The bodies are mute, unable to communicate except through snapping their fingers, but as other audience members arrive in the lobby, mimed interactions are possible and plentiful. The lobby is deserted, sand from the desert outside the theatre invading through every crevice. The audience can move through the space, and as they do, glowing bottles begin to appear. If a participant chooses to pick up the bottle, they catch a glimpse of a different reality parallel to the lobby. A lush forest appears, for example, where there was once only a wall. The physical body of the participant doesn't possess eyes that could show them this other world, but the virtual body of the participant is beyond the biological body. In planning for this moment, Piehole, the theatre company behind the production, demonstrates to the audience that this production is beyond what they could accomplish in a physical space.

When the performance begins, the single performer, whose body is human-like, instructs the audience to interact with the environment by throwing objects and exploring the area around space, a campfire outside of a house in the hills of Los Angeles. There are several live performers who alternated the role, and each person's own experience with the environment informs the way they share the space with the audience. From that space, the entire ensemble is transported to the world of the Tempest, where each audience member looms over Antonio's doomed ship, transforming into sprites who control the waves and the storm. These virtual bodies can do more than the Bauhaus designers dreamed of. They are not limited to the sets that were created for them; they can transform the world with their motion and bodies. In addition, this production of *Tempest* allows the world to influence the story. In one section, the ensemble is transported to Prospero's "memory room," which is filled with objects explaining the backstory of the original play. Each object holds a piece of the story, and audience members determine which ones to bring to the guide for an explanation. Not all the objects made it into the story, meaning that the way the virtual bodies interacted with the props in the environment determined how that part of the play would run.

Perhaps most intriguing is the way virtual bodies can interact with each other. During the marriage scene between Miranda and Ferdinand, two audience members played the two characters, and the performer asked them to take hands. I played Ferdinand for one performance, and as the other performer took my hand, they caressed my virtual hand with their own. My physical body did not feel the touch, but, given that I attended the performance during the novel coronavirus quarantine, the touch of a stranger took on meaning I couldn't experience in the physical world. The physical body of the other performer reached through the screen towards mine, and while I couldn't physically feel the touch, this moment points toward a time when sensors and other devices will allow us to transcend the physical limitations of the virtual world. These moments have the potential to introduce reality in ways that give the player the ability to perceive the virtual and physical spaces as real simultaneously, extending the limitations of each performance space and allowing for a hyper-immersive performance.

The Under Presents: The Tempest shows that real moments of interaction with other audience members and performers introduce a different relationship between audience and performance. While Bennett (2013, p. viii) suggests that some expectations of theatre exist "irrespective of the actual strategies of presentation" this performance and the behaviour of both performer and audience propose that something does change in the expectations of how to perform with other digital bodies within the space. The expanse of bodily possibility, therefore, suggests that future productions could inventively use this widening of the body's scope to reimagine the way theatre brings audiences into the story.

Conclusion

Virtual theatre presents a new suite of resources for performers to use to tell intriguing stories. With the possibility of lower costs for performance in VR, there could be an explosion of new work using the kinds of thrill rides, spectacular works, and immersive theatre described. Yet, while these possibilities are exciting, the principles behind the success of these early works are important. Scenography needs to create a complete experience in the virtual production. In contrast to the physical world, the designers must consider the new possibilities created with the virtual body. The connection between the user's body and those within the performance is essential for producing the intense connection of a theatrical performance. With the freedom to explore VR, audience members are no longer confined by the biological rules of the physical world. Future theatre productions in VR should explore how this freedom can heighten the storytelling experience. As haptic technology and other advances continue to advance the immersive capabilities of VR, theatre practitioners will be able to further draw audiences into theatrical performances that straddle the physical and virtual worlds, and these performances would be wise to continue pushing the boundaries of what the user's virtual body can accomplish within these performance spaces.

References

(2021) *Collider* [Online] Available at: https://www.upintheairtheatre.com/collider. (Accessed: 1 June 2021).

Alston, A. (2016) *Beyond Immersive Theatre: Aesthetics, Politics, and Productive Participation.* London: Palgrave Macmillan.

Bakchormeeboy. (2019) *Sleep No More 不眠之夜: An Interview with Punchdrunk Artistic Director Felix Barrett.* Available at: https://bakchormeeboy.com/2019/03/26/sleep-no-more%E3%80%8A%E4%B8%8D%E7%9C%A0%E4%B9%8B%E5%A4%9C%E3%80%8B-an-interview-with-punchdrunk-artistic-director-felix-barrett/ (Accessed (Accessed: 28 May 2021).

Bennett, S. (2013) *Theatre audiences: A Theory of Production and Reception.* New York: Routledge.

Bennett, T. (1995) *The Birth of the Museum.* New York: Routledge.

Borgmann, A. (1992) *Crossing the Postmodern Divide.* Chicago, IL: University of Chicago Press.

Boseman, R. and Kang, S. (2001) 'A Brief Historical Perspective on Panorama' in Boseman, R. and Kang, S. (eds.) *Panoramic Vision: Sensors, Theory, and Applications* New York: Springer, pp. 5-21.

Boswell, R. and Gardner, H. (2001) 'The Wedge VR Theatre.' *Apple University Consortium Academic and Developers Conference Proceedings.* Townsville, Australia: Apple Computer Australia, p2:p. 1--2:6.

Camp, R.Q., (2021) 'Quaint Devices: A Map of Headphone and Headset Plays.' *PAJ: A Journal of Performance and Art, 43*(2), pp. 36-50.

Carlson, M. (2013) '*The Theatre Ici*' in E. Fischer-Lichte and B. Wihstutz (eds.) *Performance and the Politics of Space.* New York: Routledge, pp. 26-41.

Ellis, M. (2008) 'Spectacles within Doors: Panoramas of London in the 1790s.' *Romanticism,* 14(2), pp. 133-148.

Fisher, J. (2018) 'Bauhaus Scenography for VR.' Virtual Creativity, 8(1), pp. 39-57.

Gennawey, Sam (2014). *The Disneyland Story: The Unofficial Guide to the Evolution of Walt Disney's Dream.* Birmingham, AL: Keen Communications.

Halprin, L. (1970) *The RSVP Cycles: Creative processes in the Human Environment.* New York: George Braziller.

Halprin, L. (2014) 'The RSVP cycles: Creative processes in the Human Environment.' *Choreographic Practices, 5*(1), pp. 39-47.

Kelly, K. (2017) *The Inevitable: Understanding the 12 Technological Forces That Will Shape Our Future.* New York: Penguin.

King, M. (1993) 'The American Theme Park: A Curious Amalgam' In Brown, R. and Ambrosetti, R. (eds.) *Continuities in Popular Culture.* Bowling Green, OH: Bowling Green State University Popular Press.

Kozel S. (2007) *Closer: Performance, Technologies, Phenomenology.* Cambridge, MA: MIT Press.

Laharia, U. (2020) 'VR Theatre Doesn't Seem So Distant Anymore.' *American Theatre.* Available at: https://www.americantheatre.org/2020/05/18/vr-theatre-doesnt-seem-so-distant-anymore/ (Accessed: 17 05 2021May 2021).

Martin, C. (2013) *Theatre of the Real.* New York: Springer.

Merleau-Ponty, M. (2013) *Phenomenology of Perception.* New York: Routledge.

Nilson, S. (2011) 'America's Salesman' in Riper, A. (ed.) *Learning from Mickey, Donald and Walt: Essays on Disney's Edutaintment Films.* Jefferson, NC: McFarland and Company, pp. 237-254.

Oddey, A. and White, C. eds. (2006) *The Potentials of Spaces: The Theory and Practice of Scenography and performance.* London: Intellect Books.

Piccolin, L. (2007) "Rundumkinos. Vom Panorama zu 360°-Filmsystemen." *Medienwissenschaf* 78, pp. 1-15.

Raynor, A. (1994) *To Act, To Do, To Perform: Drama and the Phenomenology of Action.* Ann Arbor, MI: University of Michigan Press.

Seamon, D. (2018) 'Architecture and Phenomenology' In Duanfang, L. (ed.) *The Routledge Companion to Contemporary Architectural History.* New York: Routledge, pp. 127-145.

Shuangjie, S. (2016) 'New York's Popular Immersive Theatre Production Coming to Shanghai.' *Global Times* 20 7 [online]. Available at: https://www.globaltimes.cn/content/995432.shtml (Accessed (Accessed: 30 May 2021).

Sternberger, D. and Neugroschel, J. (1977) *Panorama of the 19th Century. October* 4, pp. 3-20.

Walt Disney Corporation (1984). *The Walt Disney World of Entertainment,* Burbank, CA: Walt Disney Publications.

White, G. (2012) 'On Immersive Theatre.' *Theatre Research International* 37, pp. 221-235.

Section 4: VR and exhibitions

Chapter 10

Curating for communal engagement in VR

Julia Salles

Université de Montréal

Jessie Marchessault

Independent Scholar

Bart Simon

Concordia University

Abstract

Following the emergence of commercial VR headsets, a variety of venues have begun showcasing VR content in public spaces. In this chapter, we aim to address some of the challenges faced by curators exhibiting VR pieces in these settings. As such, we explore tactics to stimulate audience engagement in the pre-headset moment, while also examining how a moderator can aid immersion in VR. We illustrate these points by presenting a case study during a partnered event. This approach has allowed us to explore different curatorial strategies for creating communal forms of engagement in experiencing VR in public spaces.

Keywords: VR; curation; public space; exhibition

***

Introduction

In 2015, the Phi Centre, a multidisciplinary arts and culture organisation in old Montreal, hosted its first major VR installation, *Sensory Stories,* an exhibition that brought sixteen different cinematic and interactive VR experiences into the space and received over 12,000 visitors (Kelly, 2016). The program included a diverse range of works, such as Montreal-based studio Felix and Paul's documentary *Herders,* interactive fiction *Possibilia,* by the duo Daniels, where viewers decide how a breakup should happen, and the immersive full-body flight simulator *Birdly,* which provided a bird's eye view of Manhattan. On the last day of the exhibition, the unexpectedly long queue—which entailed queue times beyond one guest per hour—had visitors lined up around the block.

Much like the winding lines for attractions at amusement parks, crowds eager to engage with the experiences exhibited at the Phi Centre had little to do while they waited. And though this line was clear evidence of the public's growing infatuation with VR (Perry, 2016), it also revealed a key challenge of curating this media: as all of the VR pieces being exhibited were designed for single-user viewing, this meant that only one person could experience each work at a time, which also meant long queue was as inevitable as they were tedious to experience—particularly with little to no audience stimulus or preparation while waiting. Indeed, as most VR pieces are experienced alone through individual headsets, curatorial innovation is key when considering VR exhibits in public settings. In this chapter, we examine different curatorial strategies to enhance engagement and immersion in VR experiences in public spaces exploring the moments before the audience puts on the headset.

Three years after the 2015 exhibition detailed above, we interviewed Phi Centre's curator, Myriam Achard, and she commented on the many different curatorial strategies that have been tested since the *Sensory Stories* exhibition. Certainly, with the emergence of commercial VR headsets over the last six to seven years, a wide variety of venues began showcasing VR content in public contexts with different curatorial approaches to the specificities of VR, therefore, imagined. As these cultural venues and events began to integrate VR installations or projects into their programming, curatorial questions emerged surrounding how members of the public can interact with these projects in unique and engaging ways. Yet, as illustrated by the long wait times for the *Sensory Stories* exhibition, venues often present little-to-no preparation for audiences as they wait to try the VR experience using the headset (or, as we will call it, the "pre-headset moment"). Moreover, the issue of individual versus collective viewing experiences has become a challenge for curators when it comes to VR, as it is usually experienced individually, with only one person capable of wearing a head-mounted display (HMD) at any one time. While there have been several experiments that recently tackled these creative challenges (Klimek, 2018; Gifreu Castells, 2020; Neddermeyer, 2018; Allen, 2017), more research is still needed in the context of the curation of VR.

Moving forward, then, the example detailed above serves as a reminder that the physical environment where the VR experiences take place cannot be ignored. As we contend in this chapter, immersion does not happen exclusively in the headset. Even if VR creates a fully immersive environment for the audience, the ways that the digital content is physically presented in the space, including the settings, lighting, sounds, and presence of other people, have an important impact on the experience itself. Nonetheless, curation of VR experiences often focuses on the "headset moment," but it is just as important to curate the experience outside and around the headset. We would, therefore, argue that the full VR experience includes the moments just before and after the time spent in the headset, as well as the transition from physical to VR (and back again). Moreover, as Atherton and Wang argue, "[since] VR is by default an isolating medium, design for social interaction comes much less naturally than design for play" (Atherton and Wang, 2020, p. 22). Following this, strategies and principles need to be put in place to support social engagement focusing on the virtual experience (in the headset). In this chapter, then, we challenge the premise that VR is both materially and ideologically an individual experience by extending the curatorial moment to the pre-headset moment, which is always collective. In doing so, we ask the following question: how can curatorial strategies contribute to rupturing the sense that VR presents a completely immersive and individual world?

A body of work now surrounds the notion of "immersion" which is relevant to this technology. Slater and Wilbur (1997), for instance, argue that immersion is "the extent to which the computer displays are capable of delivering an inclusive, extensive, surrounding, and vivid illusion of reality to the senses of a human participant." (Slater and Wilbur, 1997, p. 603) The inclusive aspect of immersion is then defined as "the extent to which physical reality is shut out." (Slater and Wilbur, 1997, pp. 604-5). The authors also provide evidence that a higher sense of presence is reported by the audience when elements from the physical reality are incorporated into the virtual environment (Slater and Wilbur, 1997). Further still, and as Evans (2018) points out, there is a subjective and individual aspect to immersion, as the sense of presence can vary from one user to the other even when experiencing the same virtual environment. That is to say,

> immersion is far from a given, and that the ultimate goal of achieving immersion consists of combining a number of elements of different sensory stimuli and preparedness on the part of the VR user to be immersed. Immersion is reframed here as a tightly crafted emergent property of the visuals, sounds, narratives and haptics (or touch) of the VR experience and the mood or orientation of the user towards the VR experience itself. (Evans, 2018, p. 50)

In this chapter, we aim to address some of the curational questions surrounding showcasing practices and the exhibiting of VR pieces in public spaces. More specifically, we aim to address two primary challenges of VR curation in public spaces. First, how could the pre-headset moment be mobilized to contribute to the immersion of the VR experience? Second, what is the role of the moderator in a public VR showcase or exhibition? We propose the term "pre-headset" to define the moment when the user is already engaged with the VR piece she is about to experience but has not yet put the headset on. The pre-headset moment encompasses the waiting time, the lines, and the moment when the public is standing by a VR piece with some level of engagement. As literature surrounding these questions is still limited, and the research area is closely related to practices in the field, we draw our analysis from a combination of related publications, interviews with practitioners in the field of VR, and case studies examining VR exhibitions that have been curated by authors.

Curatorial strategies for VR

In their book *Rethinking Curation* (2010), Graham and Cook investigate curatorial practices in new media art and note that "the artwork's behaviour demand different modes of curating in response." (Graham and Cook, 2010, p. 283). The authors suggest that in new media art behaviour is a defining aspect for the artwork, thus curators should expand their curatorial practices integrating various modes of curation (cross-disciplinary collaboration, festivals, lab). This argument is also valid in the framework of VR. In the same perspective, Gifreu-Castells (2020) also argues that established viewing practices in visual and interactive media, such as cinema, videogames, and artistic installations, must be considered when introducing 'new' technologies into exhibition frameworks. These works bring attention to the physical and behavioural dimensions of new media, including (and maybe especially in) VR, that must be articulated to already existing curatorial practices. As Gifreu-Castells defends, it is important to consider what "differentiates the immersion of a book reader or a movie viewer from the one of VR [which] goes beyond the mind and becomes a physical action" (Gifreu-Castells, 2020, p. 362). As a result, VR draws participants into a level of physical embodiment that may disengage and/or isolate them from the contexts in which they are experiencing a project. The disconnect between a participant's physical presence within a concrete space when they put on a VR headset is a notable challenge for curators who must assure the cohesion of a VR piece and its audience within an allocated space. To address this challenge Gifreu Castells (2020) considers the following key aspects of curatorial decisions for exhibiting VR: Objectives, Theme, Technology, Space, Public, Duration, Team, and Diffusion. Gifreu-Castells then provides guidelines and recommendations for each of these aspects, arguing that it is crucial to consider the specificities of the space and technologies needed for each VR experience exhibited (Gifreu-Castells, 2020).

While guidelines can be useful to orient curatorial decisions, it is important to note that, as Caroline Klimek contends, "there are no methods for installing the different new media that automatically ensure a successful presentation" (Klimek, 2018, p. 82). In her investigation of VR programming at independent film festivals in Canada, Klimek highlights the need for "new watching habits and etiquette both in terms of interacting with and understanding this new work" (Klimek, 2018, p. 82) and calls for a more nuanced understanding of the limitations inherent in VR technologies from a programming perspective. She states that curators must approach VR projects by establishing the presence of an onsite 'mediator' or 'moderator' and suggests that both their presence in both pre-and-post moments can help ease viewers into VR experiences, creating more communal forms of experiencing creative VR works. Another notable approach to VR curation identified by Ina Neddermeyer (2018) in her analysis of current VR curatorial trends and experiments poses the question of how curation can entirely transform an existing VR artwork (for the better or worse), making curation a creative act in itself.

Drawing from literature (Allen, 2017; Gifreu-Castells, 2020; Neddermeyer, 2018; Klimek, 2018) and interviews, we noticed two prevailing trends in the curatorial setup for VR experiences: more intimate physical space (no screens, no presence of other bodies in the surroundings, usually in a limited and private area in the public space) and more open experiences of VR, often combined with the use of screens, interactions and moderators. The first 'intimate' approach identified can be seen in many VR projects installed in gallery environments. In these configurations, there are no external screens, no presence of other bodies in the nearby surroundings, and audience members are usually brought into a private space that is darkened and/or cut off from the rest of the audience. This set-up emphasizes the individual experience as the person experiencing the project is not being watched and is enclosed in a space separate from other audience members. Moreover, the VR technology itself is often concealed and there is usually some moderation done by staff or volunteers.

The second approach has a much greater focus on the collective aspects of the experience and is often more popular for gaming conventions and film festivals. This is usually done by having a project set up in an open space where bystanders or those waiting in line can watch the person or people trying the project. This strategy is more public-facing and often incorporates large screens (accompanying visuals) that showcase the project and also include a higher level of interaction with moderators orchestrating the experience. This strategy allows audiences to watch the person in the headset and the person immersed in VR becomes a part of the spectacle. In these cases, the VR technology is usually not concealed or hidden. There are cases in which these general curatorial tendencies are sometimes combined in the same exhibition or showcase. In such cases, some of the works may be exhibited in private spaces, and others are exhibited in collective setups. It is important to note that these approaches can be mixed in the same exhibition, and they can also vary according to the genre of VR content being curated.

The challenges of presenting VR in a collective setting are also considered by VR curator and producer Catherine Allen (2017). In a report about her experience curating immersive events at the Watershed, a cultural centre in Bristol, UK, she highlights that "the audiences highly valued the VR Sessions as a group activity with the facilitated research conversations afterward playing a key part of this." (Allen, 2017, para. 31) In making VR a collective experience, she notes that the "format not only helped participants feel less self-conscious, but it also served as a jumping-off point for sharing ideas and making meaning." (Allen, 2017, para. 31) Allen's communal approach to the curation of VR removed the 'waiting' period for participants by allowing groups to engage with the works simultaneously. In effect, the communal viewing of a VR experience helped comfort audience members in the pre-headset moment through the use of a pseudo-dining experience in which they were invited to 'dine' together (Allen, 2017).

The issue of individual versus collective experience has become a challenge for curators, as it is usually experienced alone through individual headsets. Elisha Karmitz, head of the now-closed MK2 VR, a venue dedicated to VR in Paris (France), stated that one of their main goals when opening was to allow for a collective experience of VR: "The idea is to offer completely interactive spaces, allowing for a real collective experience. One can see on the screens what the others are experiencing, it's a real moment of sociality and not a lonely experience. The collective experience is created by this *mise en scène* of the users" (Karmitz in Mahoudeau-Deleva, 2017, n. p., translation by the authors). As major film festivals, gallery spaces and gaming conventions now feature at least one or more VR installations or projects, questions arise surrounding how members of the public can interact with these projects in unique and engaging ways. In the next session, we present our methods, followed by the case study and the findings regarding two key aspects in the curation of VR experience in public spaces: the lack of engagement in the pre-headset moment, and the importance of the moderator in the overall VR experience.

Methods

We developed a mixed methods approach to further explore the challenges and questions surrounding VR curation in public spaces. First, we examined recent academic discussions surrounding curation, audience experience design, and VR. We then conducted in-depth semi-structured interviews with Myriam Achard, Director of Public Relations and Communications at the Phi Centre (Montreal, Canada); Samantha Cook, Producer at *Artifact 5*, a VR Game Development Studio (Montreal, Canada); Mathias Fournier, Project manager at Cinéma Contis (France); Olivia McGilchrist, VR artist and developer with extensive experience as a moderator at the Phi Centre; and Tali Goldstein, CEO & Producer at Casa Rara Studio Inc. The research questions addressed in the interviews were based on the results of the literature review. These interviews were thematically analysed, and the results informed the following phase of the research, the curation of a VR showcase in Fall 2018.

The showcase grew out of a collaboration between the Canada China International Film Festival (CCIFF) and Milieux Institute for the Arts, Culture and Technology at Concordia University. It was held at a public venue in downtown Montreal for a three-hour period before the festival's closing banquet and included two videogames and two cinematic experiences, totalling four projects that attendees could experience. We decided to include projects with varying degrees of user interactivity/playability to see what projects participants felt more comfortable with, questioning whether the presence of a moderator was more necessary when a project required audiences to engage with both head motion and controllers. The two videogames included were *Anamorphine* (Artifact 5, 2018) and *We Happy Few: Uncle Jack Live* (Signal Space Lab, 2018). The two cinematic experiences we featured were *Space Explorers* (Felix and Paul Studio, 2018) and *Mel's Wake* (Signal Space Lab, 2018).

To explore our questions surrounding how to engage audience members before they try out a VR experience (the pre-headset moment), we tested different strategies with some of the projects. First, we installed a large screen above Signal Space Lab's *We Happy Few* which audience members passing by and/or waiting in line to try the experience could watch. The screen featured a promotional teaser trailer of the project as well as short snippets of gameplay that played on a loop showing what the person currently wearing the headset was seeing (VR point of view).

In the case of *Anamorphine*, we created a three-step progression to immerse waiting (pre-headset) audience members into the game. First, we used the game's soundtrack which played in the general vicinity of the project, then we offered participants the opportunity to try the 2D non-VR version of the game (on PC) while they waited, and finally, we offered the VR experience with a monitor displaying what the person playing the VR version was seeing. By setting up this three-step process, we hoped to determine whether people would navigate from the non-VR version of the game to the VR version, all the while having the music 'set the mood.' Unfortunately, the environment was not appropriate for testing the soundtrack as the first step towards immersion as there was far too much background noise (chatter) to hear the music properly. Our analysis is therefore based on the transition from the non-VR version of the game to the VR version.

To assess the significance of having a moderator, we set up one project (*Space Explorers*) to be left unattended throughout the exhibition and another (*Mel's Wake*) to be fully attended to by a moderator. Because *Space Explorers* required little interaction on the part of the viewer, we felt it would still be accessible for passers-by and had it set up on an Oculus Go (which is wireless and very lightweight).

Throughout the 3-hour exhibition, we conducted observational research and held semi-structured interviews with six festival attendees before and after they tried the experiences. We also collected data by videoing the whole 3-hour exhibition and photographing the setup of each project. Doing so allowed us to tackle the challenges and questions surrounding exhibiting VR in public spaces by employing curatorial strategies to stimulate audience engagement in the pre-headset moment, focusing especially on the tensions between individual and collective viewing experiences. This was followed by a post-mortem session where we compared observations and notes. In the following section, then, we outline different curatorial strategies for VR, before presenting the findings from our case studies.

Pre-headset engagement and the role of moderators

As described in the introduction to this chapter, the Phi Centre's first major VR installation (*Sensory Stories*) created problems due to long wait times and a lack of pre-headset engagement. This challenge was subsequently addressed in the interview with Myriam Achard, director of Communications and Curation at Montreal's Phi Centre. While reflecting on the installation, Achard (2018) stated that though this line did indeed reflect growing public interest for the new medium, it also highlighted a key challenge of curating this media. Achard's story ties into the questions of individual versus collective experience and spectatorship which is undoubtedly a challenge for curators when it comes to VR, as it is usually experienced alone through individual headsets—especially when thinking about how VR can be experienced in public settings. If VR is so closely connected to immersion shouldn't the pre-headset/waiting period be used to prepare and/or ease participants into the environments and narratives they are about to experience? We believe that the pre-headset moment is full of curatorial potential, especially when thinking about how audiences can be involved in both pre- and post-headset moments.

We have also examined how and if the presence of a human moderator can contribute to the overall experience of a VR project. Can the moderator provide a better transition between the pre-headset and headset moment in public contexts? Further, how can the moderator help contextualize a VR experience by acting as a knowledge mobilizer that provides information, instructions and background to the project both prior and during these experiences? The thread overarching these questions is the necessity of considering the impacts and potentials for audience engagement in the physical aspects of curating VR experiences. With immersive media, it is all too easy to assume that the physical space where one is situated when experiencing VR is a less important factor for curators and that their focus should rest on the content.

Case study: Pre-headset engagement

In our study, we found that by setting up different media related to the VR experience (such as 2D monitors or audio) to be seen or heard before the actual headset experience, we contributed to contextualizing the work and instating a progressive immersion. We observed that the monitor mirroring the VR experiences of both *Anamorphine* and *We Happy Few: Uncle Jack Live* especially grabbed a lot of attention from passers-by. People gathered to see the mirrored versions of the VR experiences, thus triggering engagement. There was a lot of hesitation around trying the pieces at first (no one wanted to go first) with a lot of people looking at the setup and taking photos/videos of the 'booths' but not engaging. Moreover, in the case of *Anamorphine*, the social aspect of playing in public changed the experience, it became a more "fun" and less intimate experience. The audience gathered around the players and engaged with the players even though they were wearing VR headsets. The variety of formats for *Anamorphine* (non-VR and VR) also favoured reaching different audiences, bringing an audience of non-gamers to VR.

In the case of *We Happy Few*, we found that the flat screens worked well to grab the attention of people passing by, who watched the screens while they waited. The presence of screens also helped people get a sense of what they were about to experience, allowing them to gauge whether or not they wanted to try the project. However, while the screens showing the VR experience grabbed the audience's attention, they did not necessarily translate to participants wanting to try the experiences. There also wasn't much movement from non-VR to the VR versions of *Anamorphine*. The non-VR version reached a different audience, but once audience members had tried the non-VR version, they often felt as if they had 'seen' enough and very few gravitated to the VR version. Moreover, for *Anamorphine*, the social aspect of playing in public (with many people watching and reacting to the person playing) changed the experience, it became a more "fun" and less intimate experience.

These findings concerning the physical display of content related to the VR experience corroborate Phi Centre's curator Myriam Achard reflections on the importance of contextualizing VR experiences:

> [I]f somebody wants to open a space, or gallery, museum, that will showcase mainly XR, this person needs to know that, for instance, just having a few chairs and a few headsets and no contextualisation won't work. Because if people want to have an experience on the VR headset, they can do that from home. What a physical space needs to bring is the contextualisation of the works. (Achard in TechvangArt, 2021, para. 49).

While contextualisation is a common practice at art galleries, it is significantly less present at larger cultural or gaming conventions presenting immersive content.

We also found that the pre-headset moment can be mobilized to make the audience more comfortable with the technology. As Mathias Fournier, curator at Cinéma Contis (France), notes in one of our interviews (Fournier, 2020), it is important to demystify VR for a film audience. Fournier narrates that the discomfort of the audience is translated in a comment he hears often: "VR is not for me, I'm not a gamer" (Fournier, 2020). Indeed, one of the biggest challenges for film festival curators, as we found in our interviews, is making the film audience relate and feel comfortable trying out VR experiences. As handling the technology is already one element that creates tension among the audience, it is important to create a familiar experience in VR by making it as close as possible to the cinema-going experience. As Fournier argues, "it was important that they go buy a ticket, as they usually do for a film" (Fournier, 2020, translation by the authors). This is often done by creating dedicated space for festival goers to engage with VR pieces. For Fournier, these pre-headset moment strategies allowed creating a VR experience design that related to the film audience thus making VR experiences more accessible for non-gamer audiences.

Case study: Role of the moderator

Our findings show that many of the curatorial decisions about the physical settings and social interactions interfere greatly in the audience engagement with the virtual experiences, and decisions concerning the preparation and presence of human moderators are of great importance. Indeed, in our interview with Samantha Cook (2018), one of the co-founders of an indie game studio in Montreal called Artifact 5 which launched a narrative VR videogame in 2018, she highlighted a major issue that occurred when there was no moderator/mediator present at one of Artifact 5's VR showcases. She stated that at MEGA (Montreal Expo Gaming Arcade) in 2018, their VR game *Anamorphine* was on display a corner without a moderator and as a result, very few people felt comfortable enough to try the game. If they did try it, "they would come and go very quickly" as there was no one there to explain the mechanics and premise of the game. While this issue is more pertinent to larger exhibition spaces, Samantha Cook also described a variety of other 'expositions' where their project was similarly set in an easily accessible and obvious location at the event yet very few people ended up interacting with the project because there was little to no moderation provided by the organizers (Cook, 2018). Without a moderator on site, she believed people were too intimidated by the headset to approach the project on their own. Moreover, because of the interactive nature of the project (videogame) attendees who are not used to using the VR controller were unlikely to test the project without any help.

In our case study, while the moderator of *Mel's Wake* invited passers-by to try the experience, leading to faster engagement with the project, it took some time for people to begin interacting with the headsets displaying *Space Explorers* (left without moderators). Yet interestingly, once people saw others engaging with the piece, they appeared more likely to do the same. However, when asked, every participant who engaged with *Space Explorers* said they would prefer if there was a moderator. One female, 40-year-old participant mentioned, "I don't know how much time I can take to figure out the gameplay. In public, I don't know if there are people waiting, so I prefer when a moderator tells me how to advance in the game/story." Furthermore, it was the VR experience that was the least interacted with of all four projects. This illustrates that without some form of moderation, people are less likely to engage with VR pieces.

We also noted that human moderation of VR projects created comfort and confidence for those who are less familiar with the equipment and VR. When an audience member mentioned feeling nervous about the experience before *Mel's Wake*, the moderator reassured her and helped her adjust the headset so it would fit over her glasses. This created a sense of comfort for the participant before and after the experience as she felt that the moderator was 'looking out for her' and would be there to help her if anything went wrong.

These findings reaffirm what curator Myiriam Achard emphasizes regarding the important role of the moderators for public VR experiences: "mediation is highly important, for spaces that are presenting the XR [and VR projects]. We have mediators that make a difference in people's experiences- they know that onboarding and offboarding are as much important as the experience itself, almost. So, this is something that I would really recommend being a priority" (Achard in TechvangArt, 2021, para. 50). The physical presence of a trained moderator who can act as a mediator between the pre-and-post headset moments is critical — this individual (or individuals) not only help participants with the VR technology but can also prepare them for the kind of content they will be experiencing and can consult (or check-in) with them after the experience. Indeed Klimek's (2018) contention that patrons unfamiliar with new media are dependent on a 'translator' (or moderator) to provide "instructions, demonstrations, and tours — similar to gallery attendants in that they provided their own interpretations of the new media work." Details on how to interact with and use each device are critical to projects that require some interaction through the use of controllers or head movements during the experience.

What we've learned and where do we go from here?

Our research and case studies have helped us conclude that both the pre-headset moment and the moderator are highly important considerations to take into account while curating any kind of VR experience or showcase. More generally, we observed that decisions about the physical settings and social interactions when exhibiting VR in public venues are important factors to be considered by curators. Alongside the content, it is important that curators consider how and where the VR piece is presented, as experience outside (pre-and-post) headset is just as important as in the headset.

We have found that while established viewing practices must be considered when incorporating these technologies into spaces typically populated by visual and interactive media such as cinema, videogames, and artistic installations, the concrete space around such experiences is equally as important as the experiences themselves. Indeed, as viewers prepare themselves to engage with VR pieces, the context of the piece should be reflected in the atmosphere where it is displayed. Furthermore, the moderator should act as much more than just a headset dispenser by providing guidance, aid, and reassurance, all while preparing the viewer for the content they are about to experience.

We summarize our findings in a list of 5 curatorial lessons for exhibiting VR in public spaces:

1. Look for ways to break the solitary and individualistic sensibility around VR experiences. As we have observed in our tests, using the 2d version of the game *Anamorphine* helped to contextualize the experience of the VR game while giving audiences an opportunity to engage with one another around and across the objects of their experience. Thus, bringing VR into public spaces creates the possibility for collective engagement with immersive experiences.
2. Use the pre-headset moment to help contextualize the experience. We noted that the moments usually spent by the audience waiting for their turn to use the VR headset could also be designed to contextualize and prepare the audience to transition into VR. In the setup designed for the game *We Happy Few*, we have observed positive results in different curatorial approaches to contextualizing the VR experience, such as the use of 2D screens showing trailers and promotional video of the VR content and/or a real-time projection of the player's immersive experience.
3. Make sure of attendants to frame the experience and engage audiences before the headset phase. The transition from physical space into VR can be highly improved if the audience feels accompanied by a

moderator or facilitator, as we have noted in the curatorial tests using the VR experience *Space Explorers*.

4. Find ways for viewers to interact with one another - before, during, and after the experiences. We observed that creating occasions for interpersonal relations among the audience is a curatorial strategy with the potential to increase engagement with the VR experience, as in the three-step progression to immerse waiting audience members into the game *Anamorphine*.
5. Mix VR experiences of different duration and technological affordances. There is no clear standard VR format (as with film or games), and this could be an opportunity for curators to explore different formats and strategies to engage the audience and accentuate immersion in the VR experience.

While these elements can create opportunities for participants to feel more engaged, leading to a more 'immersed' experience, further research on the questions regarding curatorial strategies to create more collective VR experiences is still needed.

References

Achard, M. (2018). *Curation for VR*. Interview by Julia Salles, 29 May 2018.

Allen, C. (2017). 'VR Sessions: Lessons learnt for the cultural cinema sector Pervasive Media Studio.' [online] *Watershed*. Available at: https://www.watershed.co.uk/studio/news/2017/09/18/virtual-reality-sessions-lessons-learnt-cultural-cinema-sector (Accessed: 15 January 2022).

Atherton, J. & Wang, G. (2020). 'Doing vs. Being: A philosophy of design for artful VR.' *Journal of New Music Research*. 49(1), pp. 35-59.

Cook, S. (2018). *Curation for VR*. Interview by Jessie Marchessault, 13 June 2018.

Evans, L. (2018). *The re-emergence of VR*. New York: Routledge.

Fournier, M. (2020). *Curation for VR*. Interview by Julia Salles [Skype], 6 February 2020.

Gifreu-Castells, A. (2020). 'Approach to the Curatorship of VR Exhibitions' in Buckley, B. and Conomos, J. (ed.) *A Companion to Curation*. Hoboken: Wiley, pp. 360-74.

Graham, B. and Cook, S. (2010). *Rethinking curating: art after new media*. Cambridge: MIT Press.

Kelly, J. (2016). 'Montreal arts centre showcases VR docs at home and abroad' [online] *CBC News*. Available at: https://www.cbc.ca/news/canada/montreal/montreal-phi-centre-1.3564135 (Accessed: 8 June 2021).

Klimek, C. (2018). 'From Programmer to Curator: How Film Festivals Are Pushing the Boundaries of New Media and Expanded Cinema.' *Canadian Journal of Film Studies*, 27(1), pp. 73-87.

Mahoudeau-Deleva, E. (2017). 'L'espace MK2VR va devenir une référence dans le monde' [online]. *Mediakwest*. Available at https://mediakwest.com/entretien

-avec-elisha-karmitz-l-espace-mk2vr-va-devenir-une-reference-dans-le-monde/ (Accessed: 14 June 2021).

Neddermeyer, I. (2018). 'I Want to See How You See. Curatorial Practices of Exhibiting VR' in Feiersinger, L., Friedrich, K. and Queisner, M. (ed.) *Image – Action – Space.* Berlin, Boston: De Gruyter, pp. 203-216.

Perry, T. S. (2016). 'VR goes social.' *IEEE Spectrum,* 53(1), pp. 56-57. https://doi.org/10.1109/MSPEC.2016.7367470

Phi Centre. (2015). *Sensory Stories at the Phi Centre.* [online] Available at: https://phi-centre.com/en/event/fost-2015-en/ (Accessed: 7 June 2021).

Slater, M. and Wilbur, S. (1997). 'A framework for immersive virtual environments (five): speculations on the role of presence in virtual environments.' *Presence: Teleoperators and Virtual Environments,* 6(6), pp. 603–616.

TechvangArt (2021). *PHI Centre: VR Art-Worlds Vision.* [online] Available at: https://techvangart.com/2021/05/06/phi-centre-vr-art-worlds-vision/ (Accessed: 7 June 2021).

Chapter 11
Exploring the phenomenology of VR-based exhibits in art museums

Eryn Parker
City University of London
Michael Saker
City University London

Abstract

The introduction of VR has the potential to offer alternative ways for visitors to interact with exhibitions in art museums, which might differ from the spatial and social norms traditionally associated with this environment. Understanding this potential is the focus of our chapter. Drawing on an original research project conducted at the Anise Gallery in London, United Kingdom, between June and July 2018, our chapter revolves around the VR-based and multisensory exhibit, *Scents of Shad Thames*. More precisely, we report on 19 semi-structured interviews with visitors who had just experienced the VR portion of this exhibit.

Keywords: Art Museums, Space, Sociality, Social Norms

***

Introduction

VR is going through a renaissance. A range of commercial headsets are now available, and the application of VR has moved beyond the games industry (Parker and Saker, 2020). Art museums, for example, are increasingly embracing VR to extend what visitors can experience (Shehade and Stylianou-Lambert, 2020). The phenomenological implications of this advancement, however, are little understood. Consequently, while established spatial and social norms readily exist in the context of the traditional art museum (Falk and Dierking, 2016), the introduction of VR has the potential to offer alternative ways for audiences to experience and interact within these environments.

Drawing on an original research project conducted at the Anise Gallery in London, United Kingdom, between June and July 2018, our chapter focuses on the VR-based and multisensory exhibit *Scents of Shad Thames*. We report on 19 semi-structured interviews with visitors who had just experienced the VR portion of this exhibit. Our chapter progresses along three lines of inquiry. First, we consider how participants interacted with the digitality of this experience, and how these interactions might diverge from a purely physical exhibit. Second, we investigate the sociability of *Scents of Shad Thames* in the context of the staff who guided and communicated with participants throughout their time in VR. Third, and finally, we investigate whether the use of VR reshaped the process of people-watching, which is a long-established behaviour within art museums.

In sum, we contend that VR impacts the spatial and social experience of art museums in the following way.

1. The use of this technology can provide visitors with the opportunity to move their bodies differently than they would were it not for VR. Importantly, this physical freedom can allow users to get closer to the digital aspects of the exhibit, which make their experience with something that feels more personal.
2. Interactions between the curator or guide and the VR user—both physical and conversational—increase and this increase can have both positive and negative effects on how visitors engage with the exhibit. While some users might understand the extra guidance as a necessary aspect of the experience, others might feel it limits their autonomy.
3. The long-established norm of people-watching can become burdensome for some visitors, leading to feelings of vulnerability as their peripheral awareness of the physical and social space of the art museum fractures with their immersion in the digital space of headset.

In the next section, we develop the theoretical framework that guides this chapter, which focuses on the spatial and social dynamics of museums. Following our framework, we describe the *Scents of Shad Thames* exhibit before outlining the methodological approach underpinning this research. Finally, we present our findings.

Physical, spatial, and social practices in art museums

The art museum experience is shaped by the physical, social and personal contexts that each visitor brings to the museum (Falk and Dierking, 2016). While museum architecture affects the way in which visitors experience exhibitions, the social context and personal motivations for going are equally important (Hornecker and Ciolfi, 2019). Likewise, space is a social product that is rooted in shared values and socially produced meanings, which are continually modified through use (de Certeau, 1984; Lefebvre, 1991 [1974]). As such, the meanings ascribed to a space, such as an art museum, change over time (Massey, 2005; Smith, 2001, p. 36). As Parker and Saker (2020) put it, "[the] art museum ... is very much a lived space in which social actions, often dependent on social norms, can alter and modify the phenomenological experience" (p. 1161).

Nonetheless, implicit rules of conduct, or social norms, govern the ways in which people behave in this setting, making social behaviour predictable and comprehensible (Barker and Wright, 1955; Falk and Dierking, 2016). Visitors, for instance, frequently behave in a manner that is predetermined by the social norms endemic of this setting. In the context of social rules, symbolic interactionism (Carter and Fuller, 2015) suggests that society is constructed through the continued interaction and negotiation between social agents and that "agreement upon the relevant norms will be entwined with the activity needed to make sense of the sort of social event to which one is party" (Edgar, 2008, p. 232). In other words, context is significant (Blumer, 1969). Social actors behave in ways that are context-specific to establish common social norms and a shared understanding of a situation (Carter and Fuller, 2015). Consequently, engaging with the art museum as a space requires certain behaviours, including (1) physical navigation of space (2) social engagement with museum staff and (3) people watching (Debenedetti, 2003, p. 54; Falk and Dierking, 2016; Pattison and Dierking, 2013; Rosenfeld, 1980). Below we explore these behaviours in more detail.

An important feature of these environments, then, is how art museums are navigated. Established norms in this context include walking instead of running and the prohibition of touching (Leahy, 2012). For Leahy (2012, p. 50), museums have developed "normative practices of spectatorship" by regulating how visitors move their bodies in these spaces, which influences how artwork is viewed and appreciated. Consequently, spectatorship is a learned technique that shapes where audiences stand, look, and how long they stay in one spot (Larkin, 2015). Additionally, exhibits and displays are often purposefully placed and designed to be experienced from a distance because people have been socialized to not touch the items contained within these spaces (Duncan, 1995; Classen, 2005; Buggeln, 2012; Diamantopoulou and Christidou, 2019; Falk and Dierking, 2016). For Diamantopoulou and Christidou (2019), compliance with this script is manifested through proper decorum and consideration of other moving bodies. The art museum, then, is a highly choreographed and performative environment with an established 'stage set and script' for visitors (Duncan, 1995, p. 10; Garoian, 2001, p. 246).

Equally, "[art] museums are ... very much shared social spaces" (Parker and Saker, 2020, p. 1162). While social interaction primarily occurs between companions

(Falk, Koran and Dierking, 1986; Falk and Dierking, 2016), visitors also engage with museum staff through structured interactions like museum tours and unscripted conversations (Antón, Camarero, Garrido, 2017). In essence, museum staff is integral to the learning experience, especially when the visitor has spent limited time planning for the visit or the specific museum experience is novel (Pattison and Dierking, 2013; Antón, Camarero, Garrido, 2017). Over time, museum staff has shifted the scope of its role from service providers to experience providers who tailor the content and depth of experiences for different visitors (Bharwani and Jauhari, 2013; Pattison and Dierking, 2013). Though research suggests that the presence of staff can increase visitor satisfaction (Anderson et al., 2003), negative experiences can occur when staff becomes teachers, resulting in a power imbalance (Pattison and Dierking, 2013). Subsequently, many visitors establish verbal or physical barriers to limit social interaction in this setting (Debenedetti, 2003, p. 58).

Finally, people-watching is a behaviour commonly associated with art museums. While the stated goal of visiting a museum is to view the artwork on display, a visitor's attention is always partially directed towards others co-inhabiting their space (Falk, 1991; Falk and Dierking, 2016; Rosenfeld, 1980). Modelling, or the process of watching others to learn how to behave appropriately, is done by visitors who observe either their own social group, other social groups, or art museum staff (Falk and Dierking, 2016). Put differently, visitors watch people in order to become oriented to the implicit conventions of their environment. Developing this point further, Casey (2003, pp. 5-6) contends that "the viewer is acutely aware of [her] place in the museum and acts appropriately. [She] positions [herself] in front of the painting in a certain way, seeing [herself] see the object, mindful of how [she] is viewed by others."

Much of this self-awareness revolves around the fear many adults have about making mistakes in front of others and the disapproval that results from deviating from established social norms in the art museum (Durkheim, 1997 [1893]).

Scents of Shad Thames and methods

Scents of Shad Thames allows visitors to travel back in time and explore the buildings of Shad Thames—buildings that were once used to store various spices and teas. Importantly, while *Scents of Shad Thames* takes place within the physical setting of the Anise Gallery, it is through the digitality of VR that visitors are able inhabit this locale. Consequently, gallery staff remained on hand throughout the experience to assist and guide visitors as and when required.

Once they put on the supplied VR headset, visitors are presented with a floating virtual map depicting the London SE1 postcode area. Six Shad Thames buildings—including Wheat Wharf, Anchor Brewhouse, Anise Building, Cinnamon

Wharf, Tea Trade Wharf, and 28 Shad Thames—then slowly emerge from the virtual map, before occupying six designated positions within the physical setting of the gallery. Importantly, each Shad Thames building corresponds with a table in the physical space. And on each table sits a crate filled with either fruit, tea or spice—depending upon the historical building in question. In short, *Scents of Shad Thames* is a multisensory experience that combines both physical and digital elements. Not only are visitors invited to ambulate the iconic Shad Thames building through the VR headset and setting, but they are also immersed in the various scents that this environment was famous for.

The goal of this study was to explore how people interpreted the Shad Thames space as a merging of the physical and digital. Consequently, we used a purposeful sampling strategy and conducted 19 semi-structured interviews with visitors who had just experienced the VR portion of *Scents of Shad Thames* between June and July 2018. Participants included 10 females and 9 males. All participants were over 18 years old and were approached within the Anise Gallery. In short, the interviews included a number of questions revolving around the physical, spatial and social aspect of the experience. The ensuing data was thematically analysed in the context of: (1) physical exploration, (2) interacting with museum staff, and (3) people watching. In the following section, we present our findings.

Physical exploration

The art museum has been regarded both as a scripted, performative stage and as a sacred place in which one is "expected to behave with a certain decorum" (Duncan, 1995, p. 476; see also Diamantopoulou and Christidou, 2019; Leahy, 2012;). In this vein, visitors often move their bodies in a choreographed fashion framed by museum guidelines and social prompts. However, those norms are complicated in VR exhibits, and our research found that participants at the Anise Gallery explored the VR exhibit in a manner that encouraged closer examination of the artwork in a manner that was less stilted. This point can be observed in the following quotes (all names have been changed):

> VR gives you freedom to explore where you want to look, it gives you freedom to walk about. You're directing your own path (Lewis, male)

> I suppose I did actually get closer to it for some reason. I probably stepped nearer than just standing back and looking because you felt that you were so close to those buildings that you could touch them really. (Laura, female)

> Well, here I feel like I can get closer to the image. Museums and art galleries, you can't. So, I think that's the difference. And when you know it's like VR, you can put your hands and move freely, and it gives you quite more freedom to do it. (Max, male)

> But I think the special thing with this, this specific experience, is that you can approach the piece in this case. You can look around. You can get closer. (Jo, female)

Unless invited to do so, visitors of traditional museums do not usually interact or get too near to this or that display (Leahy, 2012). Indeed, it is customary for visitors to view artworks from a distance enforced by physical restrictions, like rope barriers or alarms, and by expectations set by social norms (Duncan, 1995, Classen, 2005; 2005; Buggeln, 2012; Diamantopoulou and Christidou, 2019). Yet, the extracts above indicate that many of our participants had a desire to get closer to this exhibit, which challenged the otherwise highly "choreographed environment" of the gallery (Garoian, 2001, p. 246).

Developing the point further, for some participants the importance of this freedom was the transformative effect it had on their perspective:

> It actually gave you that ability to see differently. The virtual in VR makes you feel the size of the buildings, or you know, you could be looking up and down. (Laura, female)

> It becomes more fun because you've stepped into the art. You're stepping into an art piece. (Emily, female)

> I've seen gorgeous photographs, I've seen great prints, and I've had interesting conversations, but as far, I've gone into a different world for ten minutes and that's something. (Jaz, female)

> I think it's just in the way that you can feel it more close to you. When you have the ability to touch something or feel it very near or very close, or to talk to the artist or those kinds of things, you feel more engaged with the art. (Jo, female)

Having effectively "gone into a different world" and stepped "into an art piece," participants were able to inhabit or become the artwork in a manner that would not be possible were it not in VR. The digitality of the setting, therefore, revealed something that would otherwise remain concealed in the context of a more traditional exhibition (Berkman and Akan, 2018). In other words, for our participants the ability to be closer to the artwork produced an experience that felt notably more engaging.

Our interviews also found that the use of VR effectively established new visual perspectives that produced feelings of playfulness and liberation not typically associated with art museums. This point is supported by the following quotes:

> You're just more interactive with what you're seeing so you, like it's easy to look at pictures on a wall but when you're asked to do things or to

> interact with them, that immediately becomes more, it is more fun, it's more interesting and more memorable (Emily, female)
>
> certainly, walking through the model was a strange thing to do. Because, you know, architectural models, you tend to keep a fairly respectful distance from because they're very expensive things. So, to walk through this model was like whoa, that's very strange. (Ade, male)
>
> You didn't see any other people moving, walking, talking. You just played around in the space and you were enjoying it. (Lewis, male)

As Elden (2004, p. 192) explains, "the meaning of the space and the space itself, is adapted and transformed as it is perceived and lived by social actors." For the participants above, the experience of the art museum was altered precisely because they were able to interact differently with this setting—differently than they would were it not co-constituted in VR. For Emily and Ade, the level of interactivity with the exhibit meant that a sense of transgression was felt when they breached a 'respectful distance' between themselves and the artefacts. For Lewis, however, another important aspect of VR was not just the level of interactivity it enabled, but the fact it allowed him to experience an art gallery that was suddenly more playful, private and personal (Debenedetti, 2003; Diamantopoulou and Christidou, 2019).

In sum, our interview data showed that the freedom of VR complicates the spatial and social norms commonly associated with art museums; in particular, VR challenges the notion of the art museum as a sacred space in which objects should be respected through distance (Duncan, 1995).

Interacting with gallery staff

While social interactions primarily occur between acquaintances in art museums (Debenedetti, 2003; Falk and Dierking, 2016; Falk, Koran and Dierking, 1986), interactions between gallery staff were a key component of how participants experienced the Anise Gallery.

> You're often interacting with the chap, which you normally wouldn't have that direct interaction in a gallery, a person actually working through something ... it's a bit like being a blind person as well...although you have someone guiding you. (Kaz, female)

As Kaz explains, she was effectively reliant on the physical presence of a staff member to move through the virtual exhibit. Though "interaction and guidance at different points between the museum's staff and its visitors are important contributors to the experience of the visitor" (Gilmore and Rentschler, 2002), these interactions are often fleeting. Even in the context of more intensive interactions with staff members, like museum tours, the experience described by Kaz is noteworthy. For Kaz, the guidance was viewed as being a necessary aspect of the technology and its related affordances. The following extracts support this point:

> I was interacting with the gentleman that was guiding me so that I didn't bump into anything. You could totally get lost in it and be there. Having someone that isn't sharing, doesn't have the headset on, but it was somebody outside my headset telling me 'don't go there, don't bump into anything.' (Chrissy, female)

> Yeah, it was sort of reassuring that you still had a connection with the real, where you come from ... I was aware of Jacky. It was like she was my only connection to anything that wasn't visually in front of me ... If it wasn't for the gallery curator speaking to me and tugging me away as I bumped into the walls, I really wouldn't have, I would have completely gone into that other reality. (Johnny, male)

> When he's there guiding you, first he's turning you around, look at the gallery, you know the space, and when you turn around, there's a model and then there's these buildings which weren't there before I put the headset on. So, his attempt to actually guide you through it were quite successful. (Ade, male)

> You're quite happy to be instructed where to go and to put your hands out. (Emily, female)

The experience of the exhibit, then, was shared between the participant and the staff member through physical and dialogic interactions. At the same time, of course, an experiential schism persisted between these two social actors: the physical space beyond the headset was obscured for participants, just as the digital setting within the headset was obscured for gallery staff. Consequently, participants effectively inhabited a "dislocated space" where "the digital [superseded] the physical" (Saker and Frith, 2019, p. 12), and bridging this experiential gap participants required a person on the outside to observe their movements in the concrete space of the museum. As a result, gallery staff took on an additional role that extended beyond the provision of information. More precisely, this role involved limiting the degree to which the physical world impinged upon the digital experience within the headset.

Many participants viewed the physical tethering of this exhibit as being a necessary aspect of the experience. However, this was not the case for all participants. In contrast, some participants regarded the limiting of their movement through the physical display as effectively compromising their independence and, therefore, reducing their ability to explore the exhibit as they would like to:

> You did feel like you were guided around the room rather than you have your own free ability, because I knew the guy was walking me through (Pedro, male)

> The person who is guiding you, the person is kind of giving and you're just there as slightly passive ... it's all about you really and the interaction is really only one way ... But then you are kind of forced to interact with him, you're reliant on them to help guide you through. (Gemma, female)

> The apprehension with me came when I was finished, and I couldn't control me being finished. Because I had to depend on somebody taking this thing out ... I don't like having to say, hey somebody come get me out of here (Max, male)

As discussed, art museums have rules or "rhythms" of behaviour for visitors to follow and visual, ambulatory and performative practices that "have been produced via the explicit regulations of the institution, the guidance offered by museum managers and curators, and by the exemplary conduct of more sophisticated and practiced visitors" (Leahy, 2012, p. 3). Nevertheless, a relatively high level of autonomy exists as individuals choose specific things to attend to, based on their personal preferences, impacting the meaning that are associated with their museum experience. As Falk and Dierking (2016, p. 105) put it, "visitors, not the museum, are in control; they choose what to focus on." In contrast to this more typical norm, our participants were notably less in control of their ensuing experience when traversing the VR exhibit. And for some participants, this reduction in autonomy effectively reduced their pleasure in the exhibit.

The ongoing communication between staff members and participants also led to a new social practise within the art museum predicated on a certain sensory omission. As detailed above, while inhabiting the digital space of the virtual world, participants were reliant on the physical presence of staff members. However, and interestingly, staff members were not digitally present within the headsets of participants. Instead, they shared the physical, but not virtual, space. As Emily explains:

> Because you can't see him, I know you know he's there, but you can't see him, so he feels more like a sort of voice, you know, in your ear rather than he's a presence there, so that was quite nice. (Emily, female)

For Ade, the sociability of these interactions, then, effectively became an imagined aspect of the experience.

> Well, I think with Paul, Paul obviously knew it inside out, so he could imagine what I was seeing. So, in a way I was experiencing it with him because he was talking me through the aspects of it. (Ade, male)

People know how to act in certain spaces because of established social norms (Falk and Dierking, 2016). Nonetheless, this understanding shifts in the context of disruptive technologies like VR. As detailed above, participants effectively entered a 'dislocated space' that cordoned them off from their surroundings (Saker and Frith, 2019). Here, cognizance of how to act in this new setting often hinged around audible suggestions from gallery employees who were not visually present in the digitality of the headset. In other words, the VR portion of this exhibit was not a wholly dislocated experience, but one acoustically shared with accompanying members of staff.

People watching

Research indicates that people-watching is an integral and enduring aspect of the museum experience (Parker and Saker, 2020). Visitors commonly observe other visitors, both for entertainment and for the purpose of learning how to operate within the space of the museum (Falk and Dierking, 2016, p. 135). In a similar vein, our research found that people watching remained an important aspect of the experience at the Anise Gallery. The following extracts highlight this point:

> I saw this other woman doing it and she was very careful about what not to touch...that's why when I saw that girl going around very carefully, I thought probably I would need to do something like that. (Carolina, female)

> And then I saw people walking around very slowly so I thought okay, there are things to take in... There's lots of things in the space because people just like, they kept stopping at certain increments. Okay there's something there, there's something there. (Nneka, female)

> Well once you see somebody bump into the box, you think, I'll do better. (Ade, male)

In each instance, our participants effectively anticipated the phenomenology of their time in VR by reflecting on how they might act within this space. More specifically, these participants engaged in what Falk and Dierking (2016, p. 142) refer to as "modelling," where social norms and behaviours are learned by watching how other people behave in this environment.

Significantly, the inclusion of VR did not change this aspect of the museum experience. Quite the opposite. We would suggest this practice actually took on a heightened sense of importance, precisely because the exhibit differed from a more traditional experience. As a result, participants were extra aware that they too might become part of the spectacle:

> Somebody else was watching you while you were seeing a private exhibition. (Suzi, female)
>
> So, at the beginning I was like, 'oh okay, there's people watching me walking around.' (Jo, female)
>
> What I was seeing was certainly private, but how I was viewing it and how I was interacting with it, the way I was putting my hand out was public. (Ade, male)

Notably, the spatial and social norms established in the art museum bled into the virtual experience; again, the digitality of VR was not wholly "dislocated" (Saker and Frith, 2019) from the public space beyond the privacy of the device. However, for some participants, this tension meant they were not able to feel fully present in the digital space of the device.

Traditionally speaking, people-watching in art museums primarily involves "subtle, rapid glances at other families or adults" that are usually unacknowledged (Falk and Dierking, 2016, p. 135). The inclusion of VR changed this relationship for many of our participants. They told us they felt there was an increased likelihood they would be observed by those outside of their headset because their behaviours differed from the actions typically anticipated in this setting. As Johnny notes while reflecting on the experience of observing others in VR:

> It's like seeing a friend in a funny hat.' (Johnny, male)
>
> For participants like Johnny, an awareness of being watched became both a distraction and a burden while in the digital space of the technology: It's interesting to watch other people doing it before you do it because then you're aware of, well I guess, you're aware of how you will appear. (Marco, male)
>
> I think it's very in the context to feel very self-conscious about it. I'm in front of a big group of people, there's people watching, waiting to have a go on this thing. You can feel self-conscious. (Owen, male).
>
> I didn't know if someone was laughing at me when I was touching the box and nothing was there, and I was like wearing the goggles and reading the explanation from the buildings. I don't know, I felt kind of foolish...I couldn't just be there. (Suzi, female)

As demonstrated above, the experience of the exhibit was partially impaired by feelings of self-consciousness and related concerns that participants might look foolish to others. The practice of people watching in the context of VR, then, took on a negative hue for some participants, leading to feelings of vulnerability precisely because some participants believed they were deviating from established social norms in the art museum (Durkheim, 1997 [1893]).

Furthermore, concerns about how others might react to somatic transgressions were exacerbated specifically because participants were unable to observe the reactions of others to either confirm or challenge these assumptions. In addition, the imagined reactions of those beyond the headsets often superseded the immersive aspects of the technology.

Discussion and conclusion

Our data revealed at least three significant findings about VR in art museums:

1. The use of this technology provided participants with the opportunity to move their bodies differently than they otherwise would were this a purely physical exhibit. Significantly, for some participants this meant that they were effectively able to lessen the gap between themselves and artwork, which produced an experience that felt markedly more personal.
2. Interactions with gallery staff necessarily increased because participants had to be guided through the experience. For some participants, this was understood as being a required feature of the technology. For other participants, however, this was comprehended as reducing their freedom to direct the experience themselves.
3. The long-established norm of people-watching did not just persist throughout *Scents of Shad Thames* but was effectively accentuated. Here, participants actively observed other users to anticipate their own experience in VR, just as they were more conscious of being watched when they put on the accompanying device. Equally, the requirement of wearing a headset meant participants were unable to observe the countenances of other visitors, which, therefore, exacerbated their insecurities.

Obviously, this chapter is only about one group of 18 people at one specific exhibit. Consequently, we do not present these findings as generalizable to *all* museum VR exhibits because designs and visitor characteristics differ depending on the specific exhibit. However, we argue these three findings are an important early exploration of the limits and benefits of using VR within museum settings, and we hope more research is done on this area in the future, especially because we expect that museums of VR for exhibits will only become more commonplace in the coming years.

To conclude, our interviews suggest that VR has the potential to unsettle the spatial and social norms conventionally associated with art museums. For some participants, the ability to move closer to objects and artefacts—albeit virtually—led to a physical liberation of sorts based on the possibility of a different kind of relationship with the setting. At the same time, of course, not all spatial and social norms were challenged. Quite the opposite. People watching is a good case in point. Not only did this behaviour persist, but in certain instances, it was exacerbated precisely because visitors were unable to see the countenance of others beyond their headset. In other words, just because VR might offer the potential of doing things differently doesn't mean this potentiality will be automatically embraced, nor does it mean old habits will immediately be forgotten.

References

Anderson, U.S., Kelling, A.S., Pressley-Keough, R., Bloomsmith, M.A., Maple, T.L. (2003) 'Enhancing the zoo visitor's experience by public animal training and oral interpretation at an otter exhibit,' *Environment and Behavior*, 35(6), pp. 826-841. https://doi.org/10.1177%2F0013916503254746.

Antón, C., Camarero, C., Garrido, M. (2017) 'Exploring the experience value of museum visitors as a co-creation process,' *Current Issues in Tourism*, 21(12), pp. 1406-1425. https://doi.org/10.1080/13683500.2017.1373753.

Barker, R.G. and Wright, H.F. (1955) *Midwest and its Children*. New York: Harper and Row.

Berkman, M.I. and Akan, E. (2018) 'Presence and immersion in VR,' in Lee, N. (ed) *Encyclopedia of Computer Graphics and Games*. Cham: Springer. https://doi.org/10.1007/978-3-319-08234-9.

Bharwani, S. and Jauhari, V. (2013) 'An exploratory study of competencies required to co-create memorable customer experiences in the hospitality industry,' *International Journal of Contemporary Hospitality Management*, 25(6), pp. 823-843. https://doi.org/10.1108/IJCHM-05-2012-0065.

Blumer, H. (1969) *Symbolic Interactionism: Perspective and Method*. Englewood Cliffs: Prentice-Hall.

Buggeln, G.T. (2012) 'Museum space and the experience of the sacred,' *The Journal of Objects, Art and Belief* 8(1), pp. 30-50. https://doi.org/10.2752/175183412X13286288797854.

Carter, M.J. and Fuller, C. (2015) 'Symbolic interactionism. *Sociopedia.isa*. DOI: 10.1177/205684601561.

Casey, V. (2003) 'The museum effect: gazing from object to performance in the contemporary cultural-history museum,' in: *Les institutions culturelles et le numérique*. Paris: Ecole du Louvre. 8–12 September. Available at: http://www.archimuse.com/publishing/ichim03/095C.pdf (Accessed: 19 December 2018).

Classen, C. (2005) 'Touch in the museum,' in Classen, C. (ed) *The Book of Touch*. Abingdon: Routledge, pp. 275-285.

De Certeau, M.D. (1984) *The Practice of Everyday Life*. Berkeley: University of California Press.

Debenedetti, S. (2003) 'Investigating the role of companions in the art museum experience,' *International Journal of Arts Management*, 5(3), pp. 52-63. Available at: http://www.jstor.org/stable/41064797 (Accessed: 19 December 2018).

Diamantopolou, S. and Christidou, D. (2019) 'Museum encounters: a choreography of visitors' bodies in interaction,' *Museum Management and Curatorship*, 34(4), pp. 344-361. https://doi.org/10.1080/09647775.2019.1585279.

Duncan, C. (1995) Civilizing Rituals: Inside Public Art Museums. London: Routledge.

Durkheim, E. (1997 [1893]) *The Division of Labour in Society*. New York: The Free Press.

Edgar, A. (2008) 'Norms,' in Edgar, A. and Sedgwick, P. (eds) *Cultural Theory: The Key Concepts*, 2nd ed. Oxfordshire: Routledge, pp. 231-232.

Falk, J.H. (1991) 'Analysis of the behaviour of family visitors in history museums: the national museum of natural history,' *Curator*, 34(1), pp. 44-50.

Falk, J.H., and Dierking, L.D. (2016) *The Museum Experience*. Abingdon: Routledge.

Falk, J.H., Koran, J.J. and Dierking, L.D. (1986). The things of science: Assessing the learning potential of science museums. *Science Education*, 70(5), pp.503–508. doi:10.1002/sce.3730700504.

Garoian, C. (2001) 'Performing the museum,' *Studies in Art Education*, 42(3), pp. 234-248. https://doi.org/10.2307/1321039.

Gilmore, A. and Rentschler, R. (2002) 'Changes in museum management: a custodial or marketing emphasis?,' *Journal of Management Development*, 21(10), pp. 745-760. DOI: 10.1108/02621710210448020.

Hornecker, E. and Ciolfi, L. (2019) 'Human-computer interactions in museums,' *Synthesis Lectures on Human-Centred Informatics*, 12(2).

Larkin, J. (2015) 'The body in the museum,' *Journal of Conservation and Museum Studies*, 13(1). http://doi.org/10.5334/jcms.1021219.

Leahy, H.R. (2012) Museum Bodies: The Politics and Practices of Visiting and Viewing. Abingdon: Routledge.

Lefebvre, H. (1991 [1974]) *The Production of Space*, Nicholson-Smith D (Trans.). Oxford: Blackwell.

Massey, D. (2005) *For Space*. London: SAGE.

Parker, E. and Saker, M. (2020) 'Art museums and the incorporation of VR: examining the impact of VR on spatial and social norms,' *Convergence: The International Journal of Research into New Media Technologies*, 26(5-6), pp. 1159-1173. https://doi.org/10.1177/1354856519897251.

Pattison, S.A. and Dierking, L.D. (2013) 'Staff-mediated learning in museums: a social interaction perspective,' *Visitor Studies*, 16(2), pp. 117-143. DOI: 10.1080/10645578.2013.767731.

Rosenfeld, S. (1980) *Informal Learning in Zoos: Naturalistic Studies of Family Groups*. Unpublished doctoral dissertation, Berkeley: University of California.

Saker, M. and Frith, J. (2019) 'From hybrid space to dislocated space: mobile VR and a third stage of mobile media theory,' *New Media and Society*, 21, pp. 214-228. https://doi.org/10.1177/1461444818792407.

Shehade, M. and Stylianou-Lambert, T. (2020). Virtual Reality in Museums: Exploring the Experiences of Museum Professionals. *Applied Sciences*, 10(11), p.4031. doi:10.3390/app10114031.

Smith, M. (2001) 'Repetition and difference: Lefebvre, Le Corbusier and modernitys (im)moral landscape, *Ethics, Place and Environment*, 4(1), pp. 31-44.

Index

O

P

S

T

CPSIA information can be obtained
at www.ICGtesting.com
Printed in the USA
LVHW080306310822
727189LV00010B/525/J

9 781648 894749